SCHOOL
of
Assassins

SCHOOL
of
Assassins

❖

Guns, Greed, and Globalization

REVISED AND EXPANDED EDITION

Jack Nelson-Pallmeyer

ORBIS BOOKS

Maryknoll, New York 10545

Founded in 1970, Orbis Books endeavors to publish works that enlighten the mind, nourish the spirit, and challenge the conscience. The publishing arm of the Maryknoll Fathers & Brothers, Orbis seeks to explore the global dimensions of the Christian faith and mission, to invite dialogue with diverse cultures and religious traditions, and to serve the cause of reconciliation and peace. The books published reflect the views of their authors and do not represent the official position of the Society. To learn more about Maryknoll and Orbis Books, please visit our website at www.maryknoll.com.

Orbis/ISBN 1-57075-385-7

Contents

Foreword

Roy Bourgeois, M.M.
Founder, SOA Watch

On November 16, 1989, six Jesuit priests, their housekeeper, and her fifteen-year-old daughter were massacred at the Jesuit university in El Salvador. A U.S. Congressional Task Force, headed by Rep. Joseph Moakley (D-MA), investigated the massacre and reported that those responsible were trained at the U.S. Army School of the Americas (SOA) at Fort Benning, Georgia.

Months later, in a tiny apartment just outside the main gate of Fort Benning, the SOA Watch began. Through Latin American human rights reports, the Freedom of Information Act, and help from friends in Congress, the SOA Watch began to piece together the history of a school that was little known in the United States, though it was already notorious in Latin America as a *school of assassins.*

Research revealed a story that was not very complicated for U.S. citizens to understand. It is about men with guns and a combat school that has trained over 60,000 soldiers from Latin America in commando operations, psychological warfare, and counter-insurgency techniques — all paid for by the U.S. taxpayers. These soldiers then return to their home countries with the necessary skills to defend a socio-economic system that keeps a small elite very wealthy and the vast majority very poor.

Through the work of SOA Watch the stories of the six Jesuits, the four U.S. church women, Archbishop Oscar Romero, and the many others killed by SOA graduates began to be told on college campuses, in churches and work places around the country. And as these stories spread, a movement was born — a movement that was rooted in nonviolence and connected in solidarity with the suffering poor of Latin America.

It has become a tradition in the SOA Watch movement to gather at the main gate of Fort Benning every November to keep alive the memory of those killed in Latin America and to call for the school's closure. In November 1990, ten came. In November 2000, ten thousand were there. Speakers and musicians from Latin American and the U.S. have made the event into a festival of hope. And then after the memorial service at the main gate, there is the crossing of the line onto the property of Fort Benning in a solemn funeral procession carrying coffins and whites crosses bearing the names of the victims. The names are called out and, in unison, everyone responds, "Presente!" — this person is present with us today.

In November 2000 over 3,500 crossed the line and 1,700 were arrested. Twenty-six, who had crossed the line previously, were brought to trial in Columbus, Georgia. Twenty-four were sent to prison, most with six-month sentences, including nineteen-year-old college student Rachel Hayward and eighty-eight-year-old Catholic nun Sister Dorothy Hennessey.

Over the years, seventy-two women and men in the movement have gone to prison for protesting the SOA and have collectively served a total of forty-two years in federal prisons around the country. The authorities thought they could kill the movement by sending those who opposed the SOA to prison with harsh sentences and stiff fines. But the prisoners of conscience knew that the truth cannot be silenced and they spoke from prison. And every time they went to prison the movement was energized and brought more people to Georgia in November.

Due to the growing pressure put on the SOA by this large, diverse, growing movement, coupled with a 1999 vote in the House to cut funding to the school (230 for, 197 against), the Pentagon realized the SOA was in big trouble. And so, in 2000 the Pentagon came up with a plan to close the School of the Americas and change its name to the Western Hemisphere Institute for Security Cooperation (WHISC), which passed in Congress by a vote of 214–204.

The movement, of course, was not fooled. It was a new name, but the same shame. It was like taking a bottle of poison and calling it penicillin. It's still deadly. It's still about men with guns. The SOA/WHISC is still about preparing Latin American soldiers to return to their countries with the skills necessary to implement U.S. foreign policy — by any means necessary and at any cost.

Jack Nelson-Pallmeyer has given us a very important and timely book. It should be in the hands of all who hunger and struggle for justice in our world. It is our hope that Jack's book will help empower us to speak clearly and act boldly as we work to close the SOA/WHISC and change U.S. foreign policy in Latin America.

Preface

The past several years have been marked by a number of developments, including the growing strength of the movement to close the U.S. Army School of the Americas (SOA), massive anti-globalization protests worldwide, and escalation of counterinsurgency warfare in Colombia. Last of all, there has been a name change for the SOA (it is now officially known as the Western Hemisphere Institute on Security Cooperation, or WHISC). At an SOA Watch strategy meeting in Washington, D.C., in early 2001 it became clear, in light of these developments, that a new resource would be valuable. I agreed to write a book that would describe the connecting threads between the SOA, U.S. foreign policy, Colombia, the IMF-World Bank, the WTO, and corporate-led globalization, including the significance of the SOA's recent name change. My hope is that *School of Assassins: Guns, Greed, and Globalization* will inform and energize the diverse individuals and movements concerned about any and all of these important issues. Readers unfamiliar with my earlier book *School of Assassins* should find ample background on the notorious school that lurks behind the vast majority of human rights atrocities in our hemisphere over the past fifty years. Readers familiar with the earlier book will find compelling new evidence concerning the school itself as well as a much more detailed analysis of the present role of the SOA/WHISC as an instrument of U.S. foreign policy in the context of corporate-led globalization.

Many thanks to my friends at SOA Watch, who helped me in a myriad of ways as I wrote this book; to my wife Sara and children (Hannah, Audrey, Naomi), who put up with my early mornings and late nights; to students at the University of St. Thomas and elsewhere whose creative activism fills me with hope; to Robert Ellsberg at Orbis Books for his editorial help and commitment to this project; to the twenty-six SOA Watch protestors sentenced in May 2001, most for six months, for their faithful witness at the November 2000 protest at the SOA at Fort Benning; and, finally, to the late Representative Joseph Moakley who did so much to set the stage for the eventual electoral victory, still to come, in which the School of Assassins will be closed once and for all.

Introduction

The U.S. Army School of the Americas (SOA) has been an instrument of U.S. foreign policy from 1946 to the present. *School of Assassins: Guns, Greed, and Globalization* examines the shifting role of the SOA, including changes in foreign policy linked to corporate-led globalization. The book develops seven key themes.

First, the SOA changed its name in January 2001, but it is not a new or different institution. The Western Hemisphere Institute for Security Cooperation (WHISC) is the latest expression of the U.S. Army School of the Americas. The name change and present role of the school make sense in light of the strong movement to close the SOA and the changed geopolitical context. In order to communicate the basic continuity in mission and purpose I will in the chapters that follow refer often to the SOA as the SOA/WHISC.

Second, the SOA/WHISC is an instrument of U.S. foreign policy. It is not and never has been a rogue institution. SOA graduates linked to human rights abuses have not betrayed their training or mission but have been faithful servants. They have done what their superiors (U.S. foreign policy planners, SOA trainers, CIA officials, and U.S. presidents) have determined is necessary to defend U.S. interests in Latin America. For example, progressive religious workers including priests, nuns, bishops, and lay catechists have been killed in El Salvador, Guatemala, Colombia, and throughout Latin America. These deaths are not the result of misunderstandings or the unfortunate conduct of a few demented soldiers. U.S. foreign policy planners and their allies defined these people and the theology of liberation that guided them as enemies to be defeated in order to safeguard economic and strategic interests.[1]

Third, U.S. foreign policy has many consistent features but it is never static. Understanding shifts in foreign policy strategy and tactics is essential to understanding the SOA/WHISC. It would be impossible, for example, to understand U.S. foreign policy or the SOA/WHISC outside their utility in defending interests important to or determined by powerful U.S. corporations. The strategy and tactics employed in defense of these interests, however, have evolved and changed considerably from 1946 to the present. Asking readers not to treat the following dates or schema too rigidly, I will describe in the chapters that follow four stages in U.S. foreign policy.

- Stage 1 (roughly 1946 to 1979) was a period of militarization and dictatorship.

- Stage 2 (roughly 1980–90) was marked by a two-track strategy in which the U.S. intensified repression where needed (places like Central America), and otherwise utilized debt as a form of leverage, wherever possible (throughout much of the so-called third world). Institutions like the International Monetary Fund (IMF) and World Bank (WB) functioned as powerful instruments of U.S. foreign policy in Stage 2 and they continue to be important vehicles for implementing foreign policy today.

- In Stage 3 (roughly 1991 to 1997) various forms of economic power and leverage became the principal and preferred instruments of U.S. foreign policy. The successful combination of repression and structural adjustment centrally featured in Stage 2's two-track policy laid the foundation for the institutionalization of gains in Stage 3 through trade agreements like NAFTA (the North American Free Trade Agreement).

- Stage 4 (roughly 1998 to the present) is also marked by a two-track policy. Track one involves the further institutionalization of economic leverage and power as instruments of foreign policy through the World Trade Organization (WTO) and efforts to expand free trade agreements in the Americas. Stage 4 is also marked by a period of intense remilitarization linked to two factors: the instability that is an inevitable consequence of corporate-led globalization; and, the resurgent power of the U.S. military-industrial-congressional complex.[2]

Fourth, the U.S. military-industrial-congressional complex is driving many aspects of a destructive foreign policy. The collapse of the Soviet Union and the use of economic power as the preferred instrument of foreign policy created a crisis of purpose and meaning for many parts of the U.S. military-industrial-congressional complex. Groups who benefit directly from inflated military budgets need enemies and threats sufficient to justify huge expenditures essential to their wealth and power. The insecurity, power, and self-serving needs of the military-industrial-congressional complex are feeding a dangerous resurgence in militarization that has nothing to do with real security or defense needs. The destructive power of this complex is manifested in the war and ongoing sanctions against Iraq, the expansion of NATO, the "drug war," and missile defense systems.

Fifth, corporate-led globalization, contrary to the rhetoric of most corporations, many politicians, and much of the mainstream media, isn't good for everybody or even most of us. Corporate-led globalization fosters unimaginable and disastrous inequalities, threatens the environment, and undermines authentic democracy. Its mechanisms and logic are driven by economic assumptions and materialistic visions that lead human beings down disastrous pathways that can never satisfy deep longings for community, joy, spiritual wholeness, and meaning. Development expert David C. Korten writes:

> No sane person seeks a world divided between billions of excluded people living in absolute deprivation and a tiny elite guarding their wealth and luxury behind fortress walls. No one rejoices in the prospect of life in a world of collapsing social and ecological systems. Yet we continue to place human civilization and even the survival of our species at risk mainly to allow a million or so people to accumulate money beyond any conceivable need.... We are now coming to see that economic globalization has come at a heavy price. In the name of modernity we are creating dysfunctional societies that are breeding pathological behavior — violence, extreme

competitiveness, suicide, drug abuse, greed, and environmental degradation — at every hand. Such behavior is an inevitable consequence when a society fails to meet the needs of its members for social bonding, trust, affection, and a shared sacred meaning. The threefold crisis of deepening poverty, environmental destruction, and social disintegration is a manifestation of this dysfunction.[3]

Sixth, U.S. foreign policy, including its economic dimensions and military expressions, is a major factor leading the world to the brink of disaster. The arrogance and abuse of U.S. power exercised through the SOA/WHISC, free trade agreements like NAFTA, the IMF, and the WTO need to be challenged and changed if there is to be any realistic possibility for an alternative future rooted in authentic hope.

Seventh, this period of unprecedented troubles is a time of intense hope. Contradictions in U.S. foreign policy, disastrous and ongoing tragedies linked to the SOA/WHISC, and deep problems resulting from the flawed vision and destructive practices of corporate-led globalization have given rise to widespread protests among students, workers, environmentalists, and people of faith. A growing number of people are concerned about present injustices and future prospects for sustainable living. The challenge before us is to turn widespread concerns and protests into a coherent movement for social change. We need to connect issues and movements, articulate clearly and compellingly alternative visions, and develop concrete strategies for social change embodied through creative nonviolent action.

The movement to close the SOA/WHISC, together with other movements, has the potential to close a "school of assassins" and to demand the creation of an independent Truth Commission. Such a commission could name and hold U.S. leaders accountable for numerous atrocities hidden beneath layers of mythical benevolence, Cold War rhetoric, self-deception, silence, and secrecy. As an editorial in the *National Catholic Reporter* states:

Substantial documentation by human rights organizations and by the United Nations... make it clear: U.S. policies and U.S. training of troops from Central America contributed significantly to awful episodes of torture, assassination, and other human rights abuses in the region.

... Truth commissions in Guatemala, El Salvador and elsewhere in Latin America have helped those cultures to come to a certain honest, if not perfect, understanding of the horrors that occurred. We need our own truth commission and full disclosure of the CIA, military, and other government agency documents that will shed full light on our role in Central America in recent decades. As much as any of those other countries, we need to take steps to understand our role in their suffering and to seek pardon and reconciliation.

We cannot expect accountability from others around the world for acts of violence and terror if we are not willing to scrutinize ourselves.[4]

The SOA is a window through which U.S. foreign policy can be seen clearly. What is visible through this window, as we will see, shatters the myth of the benevolent superpower. This myth provides cover to corporate forces whose vision and embodiment of globalization is leading the world to the brink of economic, political, cultural, and environmental collapse. It also serves members of the military-industrial-congressional complex. This complex justifies disastrously inflated military budgets and policies that dangerously and self-servingly militarize the world.

Placing the SOA/WHISC in the context of present U.S. foreign policy goals sheds light on the "new" school. "It is now time," according to Colonel Weidner, Commandant of the SOA immediately prior to the name-change, "to move forward, restructuring, as we have in the past, to meet new needs in a new century."[5] What are these new needs? What vision and whose interests drive U.S. foreign policy? What foreign policy instruments does the United States have at its disposal in the age of globalization? What roles are assigned to the new SOA/WHISC? How does the SOA/WHISC complement other instruments of U.S. foreign policy, including other military training programs, the World Trade Organization (WTO), and the International Monetary Fund (IMF)?

If the horrific conduct of SOA graduates sheds light on both the school itself and the foreign policy that guided it from 1946 to today, then answering the above questions will take us into the heart of U.S. foreign policy. Determining answers will require us to enter and understand diverse forums and issues, from militarization and repression in Colombia to the rules and rule makers inside the WTO, from the Free Trade Area of the Americas to the Western Hemispheric Institute for Security Cooperation, from the corridors of the International Monetary Fund and World Bank to the self-serving and destructive promoters of the military-industrial-congressional complex that benefits from destructive wars in Colombia and destabilizing missile defense systems.

Finding answers to questions about the role of the SOA/WHISC and U.S. foreign policy in the context of globalization will also lead many of us into the heart of political protest and movement building. The movement to close the SOA/WHISC is one of several creative movements that challenge core economic, environmental, and political injustices that undermine the well-being of communities and ecosystems worldwide. These diverse movements address pressing issues of labor rights, sweat shops, environmental stress, growing economic inequality, deadly debt and the power of the rich to exploit it for their own purposes, and the erosion of democracy due to corporate influence and power.

A thread connecting each of these issues and concerns is a destructive, corporate-led U.S. foreign policy that must be challenged and changed. It is sobering that Colonel Weidner and many other SOA/WHISC supporters cite El Salvador as an SOA success story. The U.S. through "Plan Colombia" is primed for a similar "success" and the SOA/WHISC is playing an important role. The SOA/WHISC, other military training programs, IMF/World Bank structural adjustment, free trade agreements and the WTO, and the military-industrial-congressional complex are all links in a destructive foreign policy chain. In today's world, powerful U.S. economic forces are driving domestic *and* foreign policies that must be changed if we are to have any realistic hope of transitioning to a sustainable, healthy world.

Diverse and growing protest movements are springs of hope. They may dry up over the coming years or they may yield abundant fruit as they spill over into the desert of our present political and economic landscape. Much depends on our ability to see and communicate vital connections. Much depends on each of us joining with others and doing what we can. Collectively, our movements have the potential and the responsibility to change U.S. foreign policy and to demonstrate the power of nonviolence to fuel a politics of protest, vision, and activism that can revitalize our nation's democracy.

Chapter 1

Official History
and the People's Stories

I was in El Salvador March 24, 2000, for the commemoration of the life and death of Archbishop Oscar Romero. Twenty years had passed since the day Romero was gunned down while celebrating Mass at the Divine Providence Cancer Hospital in San Salvador. I walked through the candlelit streets of the capital that night with my twelve-year-old daughter, Hannah, a dozen students from the University of St. Thomas where I teach Justice and Peace Studies, and perhaps as many as thirty thousand Salvadorans. It is hard to convey the power of that night as we walked, mostly in silence, amidst the flickering lights of fragile flames that interrupted the darkness. Periodically, Romero's voice penetrated the night as excerpts from his recorded sermons thundered from loudspeakers attached to the back of slowly moving trucks. His prophetic words echoed through the streets that night as they had so often during his lifetime, piercing the hearts of those who loved and hated him.

Each candle's tiny flame was a beacon of courage and hope rooted in the unspoken stories and experiences of each marcher. I could only imagine the richness of their stories and the depth of their pain. Those walking streets so often filled with blood seemed determined to keep the radical Salvadoran prophet's vision alive for the sake of themselves, their communities, and their beautiful, tormented country.

Memories: The Way We Weren't
Supposed to Be

I walked that night enveloped by my own story as well as the stories
of others. I had witnessed first-hand the terrorist tactics of U.S.-
supported forces throughout Central America in the 1980s. I lived in
Nicaragua during the height of the U.S. dirty war against the people
of Nicaragua. The U.S. Central Intelligence Agency (CIA) produced a manual for the contras that encouraged them to assassinate
"government officials and sympathizers." The manual, *Psychological
Operations in Guerrilla Warfare,* included instructions on "Implicit
and Explicit Terror." Terrorizing the populace was important, the
manual said, because once the mind of a person "has been reached,
the 'political animal' has been defeated, without necessarily receiving bullets.... Our target, then, is the minds of the population,
all the population: our troops, the enemy troops, and the civilian
population."[1] "I found many of the tactics advocated in the manual to be offensive," former contra leader Edgar Chamorro testified
to the World Court of Justice. "I complained to the CIA station
chief ... and no action was ever taken in response to my complaints.
In fact," Chamorro continued, "the practices advocated in the manual
were employed by the F.D.N. [largest contra group] troops. Many
civilians were killed in cold blood. Many others were tortured, mutilated, raped, robbed, or otherwise abused."[2] According to former
CIA official John Stockwell, terrorist tactics were central to U.S.
strategy. "Encouraging techniques of raping women and executing
men and children," he said, "is a *coordinated policy of the destabilization
program.*"[3]

Chamorro testified that CIA trainers gave the contras terrorist
training manuals *and* large knives. "A commando knife [was given],
and our people, everybody wanted to have a knife like that, to kill
people, to cut their throats."[4] His affidavit to the World Court
also said:

> A major part of my job as communications officer was to work
> to improve the image of the F.D.N. forces. This was challeng-
> ing, because it was standard F.D.N. practice to kill prisoners

and suspected Sandinista collaborators. In talking with officers in the F.D.N. camps along the Honduran border, I frequently heard offhand remarks like, "Oh, I cut his throat." The CIA did not discourage such tactics. To the contrary, the Agency severely criticized me when I admitted to the press that the F.D.N. had regularly kidnapped and executed agrarian reform workers and civilians. We were told that the only way to defeat the Sandinistas was to . . . kill, kidnap, rob, and torture.[5]

The contras were remnants of a defeated U.S.-supported dictator's army. They were an army without a country, a proinsurgency force created by the United States to wage a terrorist war against the government of Nicaragua. As such, the contras could not receive formal training at the School of the Americas, although some contras had trained there previously as members of the feared Nicaraguan National Guard that kept the U.S.-installed Somoza family dictatorship in power for more than forty years.

The United States has numerous options for training foreign soldiers. The SOA was officially off limits in the case of the contras and so the United States contracted the services of Argentina's generals, including dictator and SOA graduate Leopoldo Galtieri. Galtieri headed the military junta during a bloody period in which thirty thousand people were killed or disappeared in Argentina. In 1981, CIA director William Casey arranged for Argentina's generals, experienced from a war of terror against their own people, to train the contras in Honduras. This is not the only evidence of the U.S. forging international cooperation between tacticians of terror. The chief intelligence officer for the contras (FDN) was known to have helped plan the murder of Archbishop Romero.[6]

The U.S. mission in Nicaragua, according to Casey, was to undermine a popular revolution by running the country into the ground. "It takes relatively few people and little support," Casey said, "to disrupt the internal peace and economic stability of a small country." Casey knew the U.S.-backed contras might not overthrow the Nicaraguan government, but they "will harass the government" and "waste it."[7] On another occasion Casey told the National Security Council: "We have our orders. I want the economic infrastructure hit, particularly the ports. [If the contras] can't get the job done, we'll use our own people and the Pentagon detachment. We have to get some high-visibility successes."[8] Within months the United States used its own people, known as "Unilaterally Controlled Latino Assets," to blow up an oil pipeline and oil storage tanks. "Although the F.D.N. had nothing whatsoever to do with this operation," Edgar Chamorro reported, "we were instructed by the CIA to publicly take responsibility in order to cover the CIA's involvement."[9]

The tactics advocated in the CIA manual and used by the contras in Nicaragua are similar to those featured in training materials used at the School of the Americas and by the Southern Command. Buried in a lengthy report ordered by the Intelligence Oversight Board reviewing U.S. operations in Guatemala was the following paragraph:

> Congress was also notified of the 1991 discovery by DOD [Department of Defense] that the School of the Americas and Southern Command had used improper instruction materials in training Latin American officers, including Guatemalans, from 1982 to 1991. These materials had never received proper DOD review, and certain passages appeared to condone (or could have been interpreted to condone) practices such as executions of guerrillas, extortion, physical abuse, coercion, and false imprisonment.[10]

The language of damage control ("materials had never received proper DOD review"; "could have been interpreted to condone") could not keep the cat in the bag. The Pentagon was forced by public pressure to declassify these manuals that advocated executions, torture, false arrest, blackmail, censorship, payment of bounty for murders, and other forms of physical abuse against enemies. The manual on "Terrorism and the Urban Guerrilla," for example, says that "another function of the CI [counterintelligence] agents is recommending CI targets for neutralizing," the same euphemism used in the CIA contra manual for elimination or assassination. Officers were taught to gag, bind, and blindfold suspects, skills that take on an ominous meaning when we realize that they are part of the "Interrogation" manual and that many thousands of Latin Americans were tortured and murdered during interrogation while gagged, bound, and blindfolded. The "Handling of Sources" manual notes, the "CI agent could cause the arrest or detention of the employee's [informant's] parents, imprison the employee, or give him a beating as part of the placement plan."[11]

These tactics had a long history. Many had been featured centrally in instruction materials used by the Army's Foreign Intelligence Assistance Program ("Project X") in the 1960s. These manuals were distributed widely to countries in Central and South America where the U.S. was most heavily involved in counterinsurgency programs, including Guatemala, El Salvador, Honduras, and Panama.[12] Representative Joseph Kennedy summarized the significance of the manuals. "The Pentagon," he said, "revealed what activists opposed to the school [SOA] have been alleging for years — that foreign military officers were taught to torture and murder to achieve their political objectives."[13]

Patterns of Terror

While living in Nicaragua and as a frequent visitor to El Salvador, I witnessed over a period of many years the repressive tactics used by a series of U.S.-dominated governments against people belonging to popular organizations. In El Salvador I met members of base Christian communities, several of the Jesuit priests martyred at the Catholic University (UCA), the courageous "mothers of the disappeared," labor leaders, human rights workers, and *campesinos*. I thought about many of them as I walked in the candlelit pilgrimage commemorating Romero. All were committed to justice and all were defined as enemies and targeted by U.S. foreign policy planners and their Salvadoran allies. I walked, candle in hand, fully aware that Romero and many thousands of others had been killed by graduates of the U.S. Army School of the Americas. I shed tears for all those who died, for myself, and for my country. My mourning became part of the collective sigh of hope and painful remembrance on this inspiring night.

During my first trip to El Salvador in 1983 I met with the "Mothers of the Disappeared" at a human rights office in San Salvador. These mothers, inspired by Romero and united by their grief, were determined to change El Salvador's repressive society so that their children would not have died in vain. Each had a graphic story to tell of a son or daughter taken by "death squads." Some children had "disappeared" — that is, they were taken from sidewalks, buses, schools, or homes, never to be seen again. Others were found dead, disfigured, and brutally tortured, their bodies left in garbage dumps or on street corners where they served as a public deterrent. Like a horror movie with subtitles, the women passed around photo albums of disfigured bodies and then explained the grisly scenes with words from their personal stories. Many of us wept. In a moment so powerfully poignant in its miraculous absurdity, we were served pizza by one of the mothers as another asked us to return home to change U.S. policy and then opened her blouse to reveal scar tissue where her breast had been severed by the machete of a Salvadoran soldier.

The death squads, we should remember, were members of paramilitary and military groups allied with the U.S.-trained Salvadoran security forces. They, like the contras in Nicaragua, were instruments of U.S. policy, instruments of terror. According to an article in the *National Catholic Reporter* (NCR), thousands of declassified State Department, Defense Department, and CIA documents show "that the Reagan White House was fully aware of who ran, funded, and protected the El Salvador death squads in the 1980s, and planned the 1980 death of San Salvador Archbishop Oscar Arnulfo Romero."[14]

Paramilitaries with close ties to official militaries offer a convenient cover for the repressive militaries and the political and economic interests they serve. The "human rights record" of official militaries in repressive countries like El Salvador in the 1980s and Colombia today can be "improving" alongside escalating violence carried out by paramilitaries. A similar "human rights strategy" is at play whenever the United States officially reduces military aid to a country based on human rights considerations only to have the void in training or weapons filled by Israel or, in the past, other nations such as South Korea during the dictatorships and South Africa during apartheid.

The U.S. strategy of using paramilitary groups connected to the official armed forces to terrorize civilians in the name of freedom and democracy is being repeated in Colombia. Human Rights Watch issued a report in February 2000 entitled, "The Ties That Bind: Colombia and Military-Paramilitary Links." "Human Rights Watch here presents detailed, abundant, and compelling evidence," the opening sentence of the report reads, "of continuing close ties between the Colombian Army and paramilitary groups responsible for gross human rights violations."[15]

Minnesota Senator Paul Wellstone, after two visits to Colombia, accused Gen. Martin Orlando Carreno of having a paramilitary base "operating openly under your command" near the city of Barrancabermeja. Wellstone expressed "grave reservations" about U.S. aid. During his visit, Wellstone said, "he heard repeated reports of military-paramilitary collusion throughout the country." The military "was described to me as tolerating, supporting, and actively coordinating paramilitary operations, which often ended in massacres."[16] Colombia, the country where guns, greed, and globalization intersect most graphically today, has had more soldiers train at the SOA/WHISC than any other nation.

As I held my candle that night in San Salvador, I wondered how many mothers of the disappeared were walking with me. I also thought about other Christians whose membership in base communities and faith in a liberating Gospel had led them to work for justice. I knew they were marching and I wondered how they managed to keep hope alive after the architects of the war had subjected them to so much perverse brutality. I remembered a conversation I had years earlier (1988) with a *campesino* and base community leader. He taught me about El Salvador and authentic faith as he told me about his son:

> By a miracle I am able to tell you the story of my grand crime
> for which they threatened me with death. They took my son

who was 18 years old, shot him, peeled off his skin and cut him into pieces. Then they hung him from a cross in a tree. They cut his testicles off and put them in his mouth. They did this to warn me because I was a celebrator of the word of God. That was my crime.... We had to leave because they persecuted the whole land. Our crime is to be poor and ask for bread. Here the laws only favor the rich. However, the great majority of people are poor. Those who have jobs are exploited daily in the factories and on the farms. Without land we cannot plant. There is no work. This brings more hunger, more misery. We are without clothes, schools or jobs. And so we demonstrate. But to speak of justice is to be called a communist, to ask for bread is subversive. It is a war of extermination.... It is a crime to be a Christian and to demand justice.

Salvador:
An SOA Success Story?

As I remembered this story of the *campesino's* crucified son and hundreds of others like it, I thought of the widespread role of the SOA and its graduates in persecuting progressive religious. I also thought about the debate I had at the Cleveland City Club with the school's commandant, Colonel Glen Weidner, in February 2000, a month before Romero's commemoration. Colonel Weidner lifted up El Salvador as an SOA success story. He cited El Salvador's movement from brutality in the 1980s to democracy in the 1990s which he attributed to SOA training and positive U.S. engagement. He shared this perspective with a Cleveland audience that included relatives of Dorothy Kazel. Kazel, Ita Ford, Maura Clarke, and Jean Donovan, three nuns and a lay missioner, were brutally raped and murdered by SOA graduates on December 2, 1980. "No one in this uniform condones or tolerates the atrocities that were committed by the people of Latin America in the name of whatever doctrine or strategic objective they thought they were pursuing," Colonel Weidner said during the debate. "People that did that, did that despite the training they received at the SOA, not because of it."

Colonel Weidner's rosy portrait, based on the "our glorious country would never do anything like that" defense, avoids and clashes with mountains of evidence linking U.S. foreign policy and the SOA with religious persecution. The human rights group Americas Watch reported shortly after the murder of the Jesuit priests in San Salvador that despite U.S. denials religious persecution was a pattern throughout the decade of the U.S.-managed war:

> The government's hostility towards church and relief organizations was particularly pronounced: In the period November 13– December 14 (1989), there were 54 searches of 40 different church facilities and homes of church workers by Salvadoran military and security forces. Dozens of church workers received death threats and fled the country under government order or death threat, dozens more... were jailed and abused in detention, and numerous church facilities were ransacked.... The symbolic significance of the Army's murder of the country's leading academic and religious figures cannot be overstated: the deaths signal that, once again, no one is safe from Army and death squad violence.... *The Bush Administration has taken the position that the Jesuit murders were a dramatic departure from Salvadoran army policy.... In our view, the murders were entirely in keeping with Salvador's ten-year civil war....* Those responsible for almost every other instance of egregious abuse against Salvadoran citizens still enjoy absolute immunity.[17]

U.S. Drug Enforcement agent Celerino Castillo III offers compelling evidence that both the death squads and religious persecution were products of official U.S. policy:

> Lt. Col. Alberto Adame, a U.S. military advisor to El Salvador ... recommended one of his friends as a firearms instructor. ... Dr. Héctor Antonia Regalado, a San Salvador dentist, was a household name in the country's power corridors. I was shaking hands with "Dr. Death," as he was known in U.S. political circles, the man reputed to be the Salvadoran death squads' most

feared interrogator. In El Salvador, he was known simply as "El Doctor." Regalado's prestige among the right wing stemmed from his ability to extract teeth — and information — without anesthesia. I wanted no part of *El Doctor.* I asked Adame if the embassy had approved Regalado as an advisor. He said Col. James Steele, the U.S. Military Group commander in El Salvador, gave Regalado his blessing. The military obviously wanted this man aboard, human rights abuses and all…*El Doctor* harbored a boiling hatred for anything associated with Communism or revolutionaries, and showed particular disdain for the clergy, who sympathized with the peasants.

Castillo offers the following description of the cynical practices of "Dr. Death," whose superior was SOA graduate and death squad leader Roberto D'Aubuisson:

Regalado painted a vivid picture of the death squads' modus operandi. After watching their intended victims for a few days to learn their movements, a dozen men in two vans would move in for the abduction. They preferred to strike away from the victim's home, bolting through sliding doors on both sides of the van and yanking the person off the street. As the torture began, they wrote down every name their victim cried out. Regalado practiced his impromptu dentistry on the unfortunate captives with a pair of pliers. I could see these doomed, bleeding men, screaming names with faint hope their pain would end if they fed their captors enough future victims. The pain usually ended with a bullet or the edge of a blade.…Regalado was convinced the clergy were Communist infiltrators, trained in Cuba to undermine El Salvador. He considered them cowards, hiding behind the cloth as they spread their diseased doctrine to the peasants he loathed. He spoke of personally directing the deaths of several outspoken priests.[18]

Colonel Weidner's whitewash of the U.S. role in religious persecution also contradicts information I found on *the official SOA website* just days prior to our debate. "Many of the [SOA's] critics," the website said, "supported Marxism — Liberation Theology — in Latin America — *which was defeated with the assistance of the U.S. Army.*"[19] The continued use of such rhetoric should lay bare any claims that the SOA has reformed or that the WHISC is a fundamentally different institution.

Authentic reform depends on honest assessment of past crimes, true expressions of remorse, restitution, and accountability. Adam Isacson and Joy Olson of the Latin America Working Group and the Center for International Policy note in *Just the Facts:*

> The assertion that the U.S. Army defeated a brand of theology is deeply disturbing. The perception among Latin American militaries that religious and civic movements in support of the poor were equivalent to armed insurgency led to the deaths of thousands of civilians, including six Jesuit priests in El Salvador and the El Mozote massacre. Many of those responsible for these crimes were trained at the SOA. The persistence of this rhetoric indicates that at some fundamental level, those running the school have not yet grasped some very legitimate concerns about human rights and the history of the U.S. military relationship with Latin America.[20]

The SOA and Beyond

The conduct of U.S.-trained forces throughout the region reveals *a consistent pattern of U.S.-sponsored terror in defense of perceived interests.* It would be wrong, therefore, to focus attention too narrowly on the SOA. Abuses are *widespread* and are linked to various tentacles extending from a common home base within the foreign policy establishment. These tentacles include the SOA/WHISC, the CIA, the National Security Council, the Pentagon, the Southern Command, military training programs, U.S. Special Operations Forces, and many others. Because the pattern of abuse is revealed clearly in the conduct of SOA graduates, however, the movement to expose and close the SOA is considered dangerous by SOA supporters in the Pentagon, the White House, and some members of Congress.

This is the backdrop for the SOA's recent name change to the Western Hemispheric Institute for Security Cooperation (WHISC). It also explains why supporters of the SOA/WHISC speak often of the SOA's successful "Cold War mission" while refusing to discuss details concerning the abhorrent tactics used in carrying out that mission.

Chapter 2

Guns, Greed, and Globalization
Continuity and Change

The SOA is a military training school known throughout Latin America as a school of dictators, assassins, and coups. As U.S. citizens, we need to know why. Why has the SOA been implicated in every major human rights atrocity in our hemisphere in the past fifty years? Why have SOA graduates carried out a consistent, brutal war against members of the progressive churches, including the vicious murders of Archbishop Oscar Romero, the four U.S. churchwomen, and the Jesuit priests in El Salvador? And, given a shameful record of abuses, why have U.S. leaders and Pentagon officials denied wrongdoing and fought hard to keep the SOA/WHISC open as an instrument of foreign policy in the age of corporate-led globalization?

Those of us committed to closing the SOA/WHISC are rightfully shocked by the brutality of many SOA graduates. We must realize and remember, however, that their abhorrent human rights violations reflect the priorities of a reprehensible school and the foreign policy it services. It is not enough to be scandalized by the murder of Romero, the churchwomen, and the Jesuits, all killed at the hands of SOA graduates. We must have the courage to look into the heart of U.S. foreign policy and to trace the death and vulnerability of the poor to various actors connected to it.

These actors are diverse and they are assigned leading or secondary roles on the U.S. foreign policy stage depending on political, economic, and social conditions. They include military dictators and other SOA graduates; militaries and paramilitaries using repressive tactics; IMF-World Bank representatives using debt as leverage; corporate architects using free trade agreements and the WTO to impose discipline; and, representatives of the military-industrial-congressional complex, terrified of peace dividends and peace, grossly exaggerating threats, and destabilizing the world in order to defend their own privileges. We need to understand the role of the SOA/WHISC as an instrument of foreign policy at various stages during the Cold War and within the shifting geopolitical sands of the post-Cold War period dominated by the corporate architects of globalization.

Interests and Flexibility

The U.S. Army School of the Americas has been an instrument of U.S. foreign policy under several different names at several different locations from 1946 to the present. The school trains Latin American soldiers on behalf of elite clients in the United States and throughout Latin America to defend and promote economic and strategic interests in a hemisphere marked by massive injustice. U.S. foreign policy generally and the SOA specifically support policies and systems that enrich a powerful minority while leading to widespread poverty, gross inequalities, a strained environment, and constricted or nonexistent democracy. As U.S. Army General Maxwell Taylor once said succinctly, "As the leading 'have' power, we may expect to have to fight to protect our national valuables against envious 'have-nots.'"[1]

The SOA and its graduates have consistently targeted as enemies people who embody persistent hope and work for social changes that are necessary if poor majorities are to have a chance for meaningful life. Jon Sobrino, a Jesuit priest who avoided the fate of his murdered friends because he was not in El Salvador on November 16, 1989, writes:

> Wealth and power cannot exist if other people do not die, if people do not suffer in powerlessness and poverty and without dignity. . . . We say that the First World, the wealthy countries, cover up the greatest scandal in this world, which is the

world itself. The existence of two-thirds of humankind dying in poverty is covered up.[2]

Flexible tactics and weapons are employed in this ongoing and evolving war against the poor that targets "enemies" who speak out, awaken dignity, or organize alternative futures. Tactics and weapons range from terror and torture carried out by U.S.-trained militaries, paramilitaries, and death squads to structural adjustment programs and international trading rules imposed by bankers in three-piece suits. Bankers are often sent in after the torturers have created the proper conditions for "economic liberalization" and "democracy."

The SOA/WHISC is an instrument of an "any means necessary" foreign policy. It is important to underscore that differences in tactics discernable at the SOA/WHISC from 1946 to the present are rooted in shifts in the geopolitical ground. The "any means necessary" policy remains in place throughout. Shifting geopolitical dynamics help explain the school's past and present role.

During the Cold War, "any means necessary" meant repressing workers, peasants, progressive religious, students, and anyone else who challenged unjust economic systems or who called for structural changes to address the basic needs of poor majorities. Repression was carried out in the name of freedom and democracy and the fight against communism. It was a necessary part of an epic struggle against an evil empire. Today, any means necessary policies are as likely to be carried out by bankers as by soldiers or death squads. Cold War rhetoric has been replaced with globalization rhetoric. The outcome for the poor is much the same but the use of bankers and international trading rules as instruments of foreign policy make organizing alternatives more difficult yet no less pressing.

Foreign policy has been an instrument through which powerful economic groups have promoted their interests throughout our nation's history. Foreign policy planners have made choices from a diverse menu of military and economic options for decades as part of a U.S. strategy of low-intensity warfare against so-called third world peoples.[3] "Unfulfilled expectations and economic mismanagement have turned much of the developing world into a 'hothouse of conflict,' capable of spilling over and engulfing the industrial West," writes Pentagon consultant and low-intensity-conflict specialist Neil C. Livingstone. "[T]he security of the United States requires a restructuring of our warmaking capabilities, placing new emphasis on the ability to fight a succession of limited wars, and to project power into the Third World."[4]

With the escalation of the third world debt crisis in the 1980s, however, and greater economic vulnerability of the poor in the 1990s and today, economic leverage has itself become the key instrument of power projection and the key means by which the United States pursues foreign policy goals. *This partial transition from power projection through brute military force to power projection through economic leverage is the central dynamic in U.S. foreign policy in the age of corporate-led globalization.* If we are to make sense out of the historic and present role of the SOA/WHISC, then we must understand that IMF/World Bank structural adjustment programs (saps), free trade agreements like NAFTA/FTAA, and the WTO *are instruments of foreign policy.*

For decades following World War II, the United States defended economic interests and achieved foreign policy objectives primarily by supporting dictators, repressive militaries, and paramilitaries. The SOA and its graduates played and play central roles in achieving objectives through repressive means. Today, however, U.S. foreign policy is carried out primarily through the exercise of economic power over economically vulnerable nations and peoples. The goal of establishing and defending political and economic systems whose benefits flow overwhelmingly to elites is the same. The tactics are often different and flexible. Interests are defended through economic rather than military power whenever economic leverage is deemed sufficient to achieve policy goals.

Repressive Soldiers/Repressive Economics

Repressive militaries can be useful in one setting and unnecessary or counterproductive in another. Knowing the difference and implementing policies that reflect the difference are the key challenges facing U.S. foreign policy planners in the age of globalization in which the preferred means of power projection is economic.

Economic power projection offers many possibilities and advantages. It also has limitations. The military fist of U.S. foreign policy will be with us for the foreseeable future because corporate-led globalization is itself destabilizing. The transition from military to economic power is riddled with contradictions and will by necessity always be incomplete. It is hard to envision a stable world or a demilitarized foreign policy so long as nearly half the world's people, according to United Nations statistics, live and die on less than $2 a day; a world in which the "three richest people have assets that exceed the combined GDP [Gross Domestic Product] of the 48 least developed countries";[5] a world of unimaginable inequalities, environmental degradation, social and cultural breakdown, and rapid population growth, all aggravated by corporate-led globalization.

It would be a mistake, in other words, to overstate the case concerning the utility of economic power as a substitute for military power. Thomas L. Friedman, one of the most articulate and arrogant advocates of corporate-led globalization and a corporate-driven foreign policy, notes approvingly that nations have no realistic choice but to accept the dictates of what he calls the "Golden Straitjacket" and the "Electronic Herd," by which he means more or less the conditions set by the International Monetary Fund and whatever pleases international investors who can move "money around the world with the click of a mouse." The "Electronic Herd," Friedman writes, "gathers in key global financial centers, such as Wall Street, Hong Kong, London, and Frankfurt, which I call 'the Supermarkets.' The attitudes and actions of the Electronic Herd and the Supermarkets can have a huge impact on nation-states today, even to the point of triggering the downfall of governments."[6] This does not mean that a repressive military is obsolete. "The struggle for power, the pursuit of material and strategic interests . . . continue even in a world of microchips, satellite phones, and the Internet," Friedman writes. "Globalization does not end geopolitics. Let me repeat that for all the realists who read this book: *Globalization does not end geopolitics.*"[7]

Militarization, therefore, will remain an ongoing feature of the

SOA/WHISC and of U.S. foreign policy's overall approach to geopolitics for the foreseeable future. It would be impossible, however,
to understand the shifting roles of the SOA/WHISC or shifting
dynamics in U.S. foreign policy from 1946 to the present without see-
ing important changes in the economic landscape during that same
time frame. The key point is that for U.S. policy makers whether
to squash alternative movements and defend elite economic interests by sending in repressive soldiers or by imposing International
Monetary Fund structural adjustment programs is a tactical question. It depends on what is possible in any given setting. Both policy
options (or some combination of each), according to U.S. foreign
policy planners, have merit. As Friedman writes, the "United States
can destroy you by dropping bombs and the Supermarkets can destroy
you by downgrading your bonds."[8] Sending in the bankers will cause
the U.S. fewer problems with Amnesty International but doing so
doesn't always work. "Attention Kmart shoppers," Friedman writes,
"without America on duty, there will be no America online."[9] "Mc-
Donald's can't flourish without McDonnell Douglas, the designer of
the U.S. Air Force F-15."[10]

Pragmatism Then and Now

This pragmatism reflects economic dynamics in the twenty-first century but it isn't new. It reveals the "any means necessary" philosophy
at the heart of U.S. foreign policy since at least the end of World
War II. The United States, for example, supported the brutal dictatorship of Rafael Trujillo in the Dominican Republic for more
than thirty years. When the outrages of a valued dictator created
instability rather than stability, however, Trujillo had to go. President Kennedy, fearing that a revolutionary movement might succeed
in the Dominican Republic as it had in Cuba, allowed or encouraged Trujillo's assassination. In the aftermath, Kennedy reflected on
the different forms of government possible in the Dominican Republic and indicated that supporting another repressive dictator was
acceptable if other options failed. "There are three possibilities in
descending order of preference," Kennedy said, "a decent democratic

regime, a continuation of the Trujillo regime or a Castro regime. We ought to aim at the first, but we really can't renounce the second, until we are sure that we can avoid the third."[11]

Updating this pragmatism to the issue of foreign policy and the role of the SOA/WHISC in Latin America today, we can say that U.S. foreign policy planners prefer to get others to do what we want them to do by exerting economic pressure through the "Golden Straitjacket," the "Supermarkets," and the "Electronic Herd," the use of debt as leverage, or the imposition of NAFTA and WTO rules and sanctions.

A foreign policy by economic fiat is accompanied by waves of persuasive rhetoric concerning the widespread benefits of globalization, including rising living standards, respect for human rights, and flourishing democracy. If economic leverage and ideological persuasion fail, and they inevitably do in some settings, including Colombia today, then U.S. policy relies more on traditional military means such as paramilitaries, bombs, and terror. For this reason, the "new" curriculum at the SOA/WHISC remains primarily that of a combat school and it is for this reason also that the United States trains foreign soldiers in so many ways, in so many places.

U.S. training is so extensive that it is nearly impossible to monitor. Adam Isacson and Joy Olson in *Just the Facts: A Civilian's Guide to U.S. Defense and Security Assistance to Latin America and the Caribbean* write:

> Under a variety of programs and funding categories, the United States is **training** a large number of Latin American military personnel. While this should not be considered a complete figure, the authors' sources indicated that the United States trained at least 9,867 Latin American military personnel in 1998, with most training taking place within the region and not at schools in the United States. In 1998, for instance, the number of military personnel trained in Ecuador alone (over 1,200) exceeded the entire student body at the U.S. Army's School of the Americas at Fort Benning Georgia (875). The two largest facilities for training Latin American military per-

sonnel in the United States, the School of the Americas and the Inter-American Air Forces Academy at Lackland Air Force Base in Texas, trained 1,719 students in 1998. That same year, another 102 facilities in the United States trained Latin American military personnel. According to Gen. Charles Wilhelm, commander-in-chief of Southcom, 48,132 U.S. military personnel visited Latin America and the Caribbean in 1998 on 2,265 separate **deployments.** These included at least eighteen major training exercises.[12]

Critics of the SOA sometimes say that the SOA/WHISC remains open *despite* its atrocious human rights record, but it is more accurate to say that it remains open *because* its terrible abuses are integral to the school's valuable mission as an instrument of foreign policy past and present.

_____ *Chapter 3* _____

Focus on the SOA

The original location of the SOA was Panama, a nation whose very existence reflects the arrogance and abuses of U.S. power. The United States, while "demanding an Open Door in China...had insisted (with the Monroe Doctrine and many military interventions) on a Closed Door in Latin America," historian Howard Zinn writes, "that is, closed to everyone but the United States. It had engineered a revolution against Colombia and created the 'independent' state of Panama in order to build and control the Canal."[1] SOA insider and supporter Joseph C. Leuer describes events surrounding the creation of Panama and its aftermath in this way:

> In November of that year [1903], Panama declared independence from the nation of Colombia under the watchful eye of U.S. troops stationed aboard a flotilla headed by the USS Nashville moored outside the key port city of Colon. The implicit guarantee embodied in the flotilla's presence resulted in the Hay-Bunau Varilla Treaty between the new republic and the United States, allowing the U.S. wide latitude in constructing a canal through the jungles of Panama. During the next century, a U.S.-controlled Canal Zone would be created, an inter-oceanic canal would be built, and the U.S. military would establish a hemispheric presence radiating from its bases carved out of the tropical landscape.[2]

This poetic rhetoric seeks to mask the costs and consequences of U.S. militarization of Latin America and to obscure why the SOA was forced out of Panama as part of the Panama Canal Treaty negotiated in 1977 during the presidency of Jimmy Carter. The

21

Panamanians demanded the school's removal because during the thirty-some years between its opening and the Canal Treaty the SOA had developed a well-deserved reputation throughout Latin America as a school of dictators, assassins, and coups. The SOA moved to Fort Benning, Georgia, in 1985.[3] It remains there today as it continues its consistent yet evolving mission under the cover of its most recent name change to the Western Hemisphere Institute for Security Cooperation (WHISC).

The SOA functioned at the heart of U.S. foreign policy in Latin America for decades and yet few U.S. citizens knew about the school or paid attention to its mission or how Latin Americans experienced it. The fact that the SOA was known and associated with human rights atrocities, death squads, and dictatorships throughout Latin America while remaining largely invisible to U.S. citizens until the 1990s says a good deal about our nation's arrogance, citizen ignorance, and the power of militarization, secrecy, and the nation's uniting myth of the benevolent superpower.

Myth and Reality

The need to expose militarization, shatter the myth of benevolence, and take an honest look at our nation lies at the heart of an essay on Martin Luther King by Vincent Harding. I thought of this essay when I was in El Salvador for Romero's commemoration, pleased to find that Romero, unlike King, has so far avoided being reduced to a relatively benign figure. Harding argues that we need to "catch up with" King's call for a "revolution of values" and his radical critique of the United States as a militaristic nation whose military serves a destructive economic system. The radical King has been diminished and transformed into an icon by the dominant mythmakers. He "seems safely dead, now that he has been properly installed in the national pantheon," Harding writes, "to the accompaniment of military bands, with the U.S. Marine Corps chorus singing 'We Shall Overcome,' and the cadenced marching of the armed forces color guards. . . . " The real and radical King, Harding reminds us, "was voicing increasingly strong opposition to the American war in

Vietnam" and he had told other audiences "he would definitely take a conscientious objector position, and would not accept a military chaplaincy." Harding continues his portrait of a radical prophet.

> King found that his search for a way to challenge the government and the nation to justice was constantly blocked by the reality of Vietnam. The poor young men of America were being swept up to become victims and executioners in ever-larger numbers. The poor of Vietnam were being destroyed physically and culturally. Moreover, King knew that all the cruel devastation of an unjust war was draining billions of dollars and lifetimes of energy and creativity out of the nation's potential for dealing with the needs of its own people.[4]

King, according to Harding, was setting out "to organize the poor for confrontation with the powers of oppression."[5] Citing King's own words, the United States was "the greatest purveyor of violence in the world today." It was "engaged in a war that seeks to turn the clock of history back and perpetuate white colonialism." The nation, King said, had to recognize that "the evils of capitalism are as real as the evils of militarism and the evils of racism" and that we had to free ourselves from "the triple evils of racism, extreme materialism, and militarism." "Something is wrong with capitalism as it now stands in the United States," King said. "We are not interested in being integrated into *this* value structure. Power must be relocated."[6] King also said:

> The storm is rising against the privileged minority of the earth, from which there is no shelter in isolation or armament. The storm will not abate until a just distribution of the fruits of the earth enables men everywhere to live in dignity and human decency.[7]

The United States, King said, was on the wrong side of the world revolution and should stand with, not against, the poor. And, as Harding notes, King predicted prophetically that without a revolution in values "people of good will in America would end up protesting our nation's new Vietnams all over the world, including

Central America."[8] Harding describes what it would mean for us to catch up with King:

> We catch up with King only as we face all the hard contradictions of the militarism he decried.... We go ahead of our brother only as we continue to call attention and to organize resistance to the militarization of our nation's budget, its children's imaginations, its foreign policy, and our lives. We catch up and go ahead by marching, picketing, sitting, undermining the ways of violence, refusing to pay taxes, giving our money to the things that make for peace, constantly searching for an answer to our brother Martin's question from long ago: Is there a nonviolent, peace-making army that can shut down the Pentagon? We go ahead by creating reservoirs of peace within us, around us, wherever we are, in whatever family and community, preparing for the coming time of great flooding, watering the small places now.[9]

In our effort to catch up to King we will undoubtedly meet Archbishop Romero. Romero, like the radical, almost forgotten King, challenged U.S. militarism and the U.S. myth of benevolence. One such challenge is found in Romero's letter to President Carter in February 1980:

> I am very concerned by the news that the government of the United States is planning to further El Salvador's arms race by sending military equipment and advisors to "train three Salvadoran battalions in logistics, communications, and intelligence." If this information... is correct, instead of favoring greater justice and peace in El Salvador, your government's contribution will undoubtedly sharpen the injustice and the repression inflicted on the organized people, whose struggle has often been for respect for their most basic human rights.
>
> The present government junta and, especially, the armed forces and security forces have unfortunately not demonstrated their capacity to resolve in practice the nation's serious political and structural problems. For the most part, they have

resorted to repressive violence, producing a total of deaths and injuries greater than under the previous military regime.... If it is true that last November "a group of six Americans was in El Salvador... providing $200,000 in gas masks and flak jackets and teaching how to use them against demonstrators," you ought to be informed that it is evident that since then the security forces, with increased personal protection and efficiency, have even more violently repressed the people, using deadly weapons.

For this reason, given that as a Salvadoran and archbishop ... I have an obligation to see that faith and justice reign in my country, I ask you, if you truly want to defend human rights: to forbid that military aid be given to the Salvadoran government; to guarantee that your government will not intervene directly or indirectly, with military, economic, diplomatic, or other pressures, in determining the destiny of the Salvadoran people. It would be unjust and deplorable for foreign powers to intervene and frustrate the Salvadoran people, to repress them and keep them from deciding autonomously the economic and political course that our nation should follow.[10]

President Carter rejected Romero's appeal and when President Reagan took office Central America became the most important place on earth. "The national security of all the Americas is at stake in Central America," President Reagan argued. "If we cannot defend ourselves there, we cannot expect to prevail elsewhere. Our credibility would collapse, our alliances would crumble and the safety of our homeland would be put in jeopardy."[11] It wasn't long before Romero and the Salvadoran people were on the receiving end of "the greatest purveyor of violence in the world." Many of the architects of the U.S. war in Indochina sought redemption of egos, careers, capitalism, and national pride in Central America.

Romero knew a good deal about the U.S. role in his country and the likely consequences of militarization. He had no way of knowing, however, that his future assassins were trained at the School of the Americas. Father Roy Bourgeois, Army reservist Linda Ventimiglia, and Larry Rosebaugh, a priest with the Oblate Order, tried to make U.S. citizens aware of the links between U.S. military training and repression in El Salvador through a provocative act of civil disobedience at Fort Benning on August 15, 1983. Father Roy has played an instrumental role in efforts to close the school. At the time, he and the others did not know Romero had been killed by SOA graduates. They did know, however, that large numbers of Salvadorans were being trained at Fort Benning while thousands of civilians were being killed in El Salvador as a result of a U.S.-sponsored dirty war. Wearing U.S.-surplus military uniforms, they entered the base carrying a boom box and a tape of Romero's final sermon, climbed a tree near the residences of the Salvadoran trainees, turned up the volume, and pressed play. Romero's voice thundered:

> I want to make a special appeal to soldiers, national guardsmen, and policemen: Brothers, each one of you is one of us. We are the same people. The *campesinos* you kill are your own brothers and sisters. When you hear the words of a man telling you to kill, remember instead the words of God, "Thou shalt not kill." No soldier is obliged to obey an order contrary to the law of God.... In the name of God, in the name of our tormented people who have suffered so much and whose laments cry out to heaven, I beseech you, I beg you, I order you in the name of God, stop the repression![12]

A day after he issued this stirring appeal to Salvadoran soldiers and less than five weeks after his letter to President Carter, Romero was assassinated by SOA graduates while celebrating Mass at the Divine Providence Cancer Hospital in San Salvador.

The spotlight on the SOA intensified as a result of Representative Joseph Moakley's investigation into the November 1989 murder of six Jesuit priests, their housekeeper and her daughter at the Catholic University in San Salvador (UCA). The official U.S. response to the Jesuit murders was swift. U.S. Secretary of Defense Richard Cheney stated emphatically, "There's no indication at all that the government of El Salvador had any involvement."[13] The Bush administration tried to pin the murder on the armed opposition while working tirelessly to keep U.S. military aid flowing. The U.S. Congressional Task Force concluded in its April 30, 1990 report that the men responsible for the massacre were trained at the School of the Americas at Fort Benning. They were part of the elite U.S.-trained Atlacatl Battalion and had recently completed a course on "human rights."

The U.S. Congressional Task Force raised concerns about the U.S. Army School of the Americas that were confirmed and deepened by the *United Nations Truth Commission Report* issued on March 15, 1993. For anyone paying attention to El Salvador in the 1980s, the Truth Commission offered few surprises. It held the U.S.-trained, -funded, and -equipped Salvadoran military and a series of U.S.-backed governments responsible for the vast majority of human rights violations, massacres, and civilian deaths, including the murders of Archbishop Romero, four U.S. churchwomen, and the Jesuits.

The Truth Commission did not address specifically the role of the United States in El Salvador. The U.S. is the elephant in a room of which everyone is aware but of which nobody speaks. The United States had provided financial (more than $6 billion), logistical, military, and ideological support to the groups the Truth Commission held responsible for most of the disappearances, repression, massacres, and other human rights abuses. The Truth Commission listed the names of Salvadoran officers most responsible for atrocities. Crosschecking the U.N. list with a list of SOA graduates revealed that *more than two-thirds of the more than sixty officers cited for the worst atrocities in El Salvador's brutal war are alumni of the School of the Americas.*[14] Keep that figure in mind when SOA supporters speak about a few bad apples or cite El Salvador as a success story.

The Jesuit murders, Congressman Moakley's Task Force, and the U.N. Truth Commission shed much needed light on a despicable institution at the heart of U.S. foreign policy in Latin America. The cat, so to speak, was out of the bag. As time went on and as the faithful protest and effective documentation of the SOA Watch office and movement catalogued startling abuses,[15] it became clear that atrocities associated with the school and its graduates were widespread. The despicable conduct associated with the school and its graduates extended well beyond El Salvador to country after country in Latin America. Collectively this evidence offers important but discomforting glimpses into a destructive foreign policy with costly domestic and international consequences.

Cold War Mission

The U.S. Army School of the Americas produced its final edition of *ADELANTE* in the fall of 2000. The "Historical Edition" of the school's official magazine opened with "A Word from the Commandant," Colonel Glen R. Weidner:

> I am proud to offer a very special edition of *ADELANTE* as the last issue to be published by the U.S. Army School of the Americas. On 15 December, 2000, the School will furl its colors after 54 years of distinguished service to the United States and to the nations of Latin America and the Caribbean. . . .
>
> Over 61,000 soldiers and civilians from 21 countries have passed through courses at the School during a period that coincided with the epic struggle of the Cold War. That conflict resonated deeply in the Americas, as externally supported civil wars aggravated traditional rivalries and the socio-economic ills that have plagued the region for centuries. As a result, the School's role in training Latin American militaries to face insurgent threats has eclipsed its broader purposes . . . to promote hemispheric peace by bringing the militaries to the region together to study professional matters, resulting in greater mutual understanding and potential cooperation. The School constituted an important piece of the system that emerged under the Charter of the Organization of American States (OAS), a system dedicated to the peaceful resolution of disputes, collective response to security threats, and social and economic progress for the peoples of the Americas.
>
> *The School of the Americas is closing, having accomplished its Cold War mission. It is now time to move forward, restructuring, as*

we have in the past, to meet new needs in a new century. The over-riding goals of the OAS Charter are still of critical importance to each of our nations, and will serve to guide the formation of a new institution dedicated to preparing military professionals to work cooperatively towards their achievement.[16]

These words reveal and distort reality in important ways. First, Colonel Weidner refers to the SOA in the past tense. As of December 15, 2000, the SOA was no more. This is "the last issue." The School furled "its colors after 54 years of distinguished service," it "constituted an important piece of the system," and it "accomplished its Cold War mission." By speaking of the SOA in the past tense, Colonel Weidner seeks to distance the SOA from the "new" school (WHISC) that replaced it.

The issue of *ADELANTE* from which this quote is taken, however, makes it clear that since 1946 the SOA has been called by different names and conducted different missions as an instrument of U.S. foreign policy. Colonel Weidner dedicates the final edition "to recounting the history of the School *in its various forms* since its founding in 1946" (emphasis added). Despite rhetoric to the contrary, in other words, Colonel Weidner is aware that WHISC is a new form of the SOA and not a new institution.[17]

A second revelation and distortion concerns the long shadow of the Cold War. According to Colonel Weidner, the "epic struggle of the Cold War" led the SOA to abandon its original purpose and to focus attention on "externally supported civil wars." The realities of the Cold War "eclipsed" the School's "broader purposes" as "the School's role in training Latin American militaries" expanded in order to deal with "insurgent threats." Like all good soldiers, the SOA did what it was told and did it well. "The School of the Americas is closing, having accomplished its Cold War mission."

One troubling feature of the name change ploy and frequent references to the school in the context of the Cold War is that it involves no real soul-searching, no acknowledgement of wrong doing, no repentance, no confession, no reparation, no sense of remorse on which an authentically new institution would need to be founded.

Joseph Leuer, by way of example, titles his article "A Half Century of Professionalism...."[18] He dismisses critics of the School and, like Colonel Weidner, cites the SOA's successful Cold War mission without specific reference to tactics employed, groups targeted, and lives destroyed:

> USARSA [U.S. Army School of the Americas] was tasked to act as the integrated training center for the warriors who fought the hot wars that spilled into the Central and South American regions as a result of the East West ideological competition of the Cold War — a mission successfully executed by the thousands of U.S. and Latin American military personnel who passed through the school, but one that ultimately led to the closure of USARSA at the culmination of the century *as a result of defamatory allegations* that it served to inspire criminal conduct by its graduates.[19]

The SOA movement is motivated, according to this view, by vengeful people with irrational concerns. More important, the Cold War mission accomplished through the SOA and its graduates is frequently cited but without examination. It seems strange to close a perfectly good school and open another with a new name at the same site with essentially the same curriculum. Colonel Weidner's desire to distance the new SOA/WHISC from the old SOA and his admission without content that the school fulfilled its Cold War mission raise two questions. Why distance a new school from the old if the conduct of the old is as noble as supporters claim? What tactics were used as the SOA fulfilled its Cold War mission?

The answer to the first question is straightforward: The "new" WHISC is distanced from the "old" SOA and a "perfectly good school" is replaced by another at the same site with nearly the same curriculum because the mission and conduct of the school and its graduates have involved horrific human rights abuses and atrocities. The name-change charade sends two prominent messages: The truth about the SOA cannot be acknowledged without compromising something of great importance to U.S. leaders; and, the human rights atrocities and other abuses linked to the SOA and its graduates were at the heart of the school's mission and helped it achieve its status as a valued foreign policy asset then and now.

Colonel Weidner and many others responsible for U.S. foreign policy speak about the Cold War mission of the SOA but they are silent about the nature of that mission and the tactics used to accomplish it. The name change is meant to obscure that many foreign policy objectives were achieved through the gruesome work of U.S.-trained SOA graduates. Weidner assumes mention of the Cold War lets him, the SOA, SOA graduates, and U.S. foreign policy planners off the hook in terms of legal and moral accountability. Anything done in the context of the Cold War and in the fight against "communism" is justifiable and therefore it is unnecessary and unwise to offer specifics. Better to maintain a deafening silence concerning details and say simply that the SOA is closing having "fulfilled its Cold War mission." The fact that Colonel Weidner places the successful mission of the SOA in the context of the Cold War invites us to look at the policies carried out by the United States under the guise of Cold War rhetoric.

Chapter 4

Evidence and Tactics

Investigation into the murder of the Jesuits and the United Nations Truth Commission focused public attention on the School of the Americas for the first time. The SOA could no longer avoid scrutiny. It was like a deer caught in the headlights of an oncoming car:

- Two of three officers cited in the assassination of Archbishop Romero are SOA graduates, including death squad founder Roberto D'Aubuisson.

- Three of five officers cited in the rape and murder of Maura Clarke, Jean Donovan, Ita Ford, and Dorothy Kazel are SOA graduates.

- Three of three officers cited in the case of two murdered labor leaders are SOA graduates.

- Ten of twelve officers cited as responsible for the massacre of more than nine hundred civilians in El Mozote, a massacre actively covered up by the U.S. government and perhaps the single most horrific event in the bloody civil war, are SOA graduates.

- Nineteen of twenty-six officers cited in the November 1989 murder of six Jesuit priests, their housekeeper, and her daughter are SOA graduates.

- Overall, more than two-thirds of the more than sixty officers cited for the worst atrocities in El Salvador's brutal war are alumni of the School of the Americas.[1]

The SOA is only one place of U.S. influence that contributed to the murder of the Jesuits and other atrocities in El Salvador. I detailed previously how U.S. drug enforcement agent Celerino Castillo III was forced by the U.S. Military Group commander in El Salvador to bring death squad leader and torturer "Dr. Death" into his employ because the U.S. "military obviously wanted this man aboard, human rights abuses and all." Dr. Death, you will remember, "was convinced the clergy were Communist infiltrators" and he spoke openly "of personally directing the deaths of several outspoken priests."

Secret documents leaked from a 1987 meeting of the Conference of American Armies reveal further U.S. involvement in repressive policies targeting progressive church leaders. Signed by military commanders from Argentina, Uruguay, Chile, Paraguay, Bolivia, Brazil, Peru, Ecuador, Colombia, Venezuela, Panama, Honduras, Guatemala, El Salvador, and the United States, the documents attack liberation theology as a tool of international communism. "The disputes brought about by this new theological reflection," the generals state, "have fostered a favorable climate and given a new tone to the Marxist penetration of Catholic — and in general, Christian — theology and practice." The documents portray liberation theology as a fundamental enemy that must be countered through a strategy of continental security measures that include the coordination of military intelligence and operations. The generals also support the use of elections as a cover for their own de facto rule. They oppose a new wave of military coups, preferring "a permanent state of military control over civilian government, while still preserving formal democracy."[2]

The Conference of American Armies named Ignacio Ellacuría, head of the UCA and one of the soon-to-be-murdered Jesuits, as a person who manipulated "the truly liberating Christian message of salvation to further the objectives of the communist revolution."[3] Little wonder Ellacuría and the other Jesuits were killed two years later by SOA graduates. Other members of the U.S.-trained Salvadoran armed forces justified and celebrated their murders. Days before the Jesuits were dragged from their beds and executed, the

Salvadoran Air Force, headed by SOA graduate Gen. Juan Rafael Bustillo, produced and distributed a leaflet saying:

> Salvadoran Patriot! You have the... right to defend your life and property. If in order to do that you must kill FMLN terrorists as well as their "internationalist" allies, do it.... Let's destroy them. Let's finish them off. With God, reason, and might, we shall conquer.

The morning after the massacre, U.S.-trained soldiers of San Salvador's First Infantry Brigade circled the offices of the Catholic Archdiocese in a military sound truck, celebrating the murders and those who had carried them out. "Ignacio Ellacuría and Ignacio Martín-Baró have already fallen," the loudspeaker shrieked. "And we will continue murdering communists."[4]

The hostile rhetoric from the Conference of American Armies report mirrors that of influential U.S. policymakers who shaped the Reagan administration's strategies in Central America. The Council for Inter-American Security, in a 1980 paper commonly referred to as the Santa Fe Report, stated:

> U.S. foreign policy must begin to counter... liberation theology as it is utilized in Latin America by the "liberation theology" clergy.... Unfortunately, Marxist Leninist forces have utilized the church as a political weapon against private property and productive capitalism by infiltrating the religious community with ideas that are less Christian than communist.[5]

The claim on the official SOA website (year 2000) that "many of the [SOA's] critics supported Marxism — Liberation Theology — in Latin America — *which was defeated with the assistance of the U.S. Army*" (emphasis added) takes on an ominous flavor in light of the pervasive hostility to liberation theology expressed in official documents and actual repression experienced by priests, nuns, and lay church workers. The evidence is clear that SOA graduates specifically, and U.S. foreign policy generally, targeted progressive religious as enemies. The question as to why will be explored in chapter 6.

The Tip of the Iceberg

The role of SOA in El Salvador is the tip of an iceberg both in terms of broader involvement of SOA graduates in the region and the foreign policy behind the school itself. Guatemala, a nation where approximately 150,000 people perished in the decades that followed a 1954 CIA-directed coup that overthrew a democratic, reformist government, offers a compelling case in point. Mons. Próspero Penados del Barrio, Archbishop Primate of Guatemala, writes in the Introduction to *Guatemala: Never Again!* the official report of the Human Rights Office of the Archdiocese of Guatemala:

> This war was characterized by torture and murder. Entire communities were obliterated, terrorized, and defenseless in the crossfire, and nature, which the indigenous cosmovision holds sacred, was destroyed. Like a maniacal windstorm, the war swept away the cream of Guatemala's intellectual community. The country was orphaned abruptly by the loss of valuable citizens whose absence is still being felt. Who was the victor in this war? We all lost. I do not believe that anyone is cynical enough to raise the flag of victory over the remains of thousands of Guatemalans — fathers, mothers, brothers and sisters, and young children — innocent of the inferno that consumed them. Our country's social fabric was devastated.[6]

The report, commonly known as REMHI (Recovery of Historical Memory Project) notes that government military and paramilitary forces were responsible for 87.38 percent of the torture cases and 89.7 percent of the other atrocities committed during the war.[7] It also details the tactics employed by military and paramilitary groups. Many of these tactics are identical or strikingly similar to those promoted in the already discussed SOA and CIA-produced training manuals that were carried out by the U.S.-trained contras and Salvadoran soldiers. Here is a sampling from the REMHI report:

- Human rights violations have been used as a strategy of social control in Guatemala. Society as a whole has been touched by fear, whether during periods of rampant and indiscriminate violence or times of more selective forms of repression. More than simply a byproduct of armed confrontation, terror has been the goal of a counterinsurgency policy that utilized different means at different times (fear is the effect most frequently reported in the testimonies (p. 4).[8]

- The strategy of forced disappearance and murders of leaders of social organizations ... [was] employed throughout the conflict. . . . The goal of selective repression has been to crush organizational efforts viewed as threatening to the government. In such cases, the police and security forces chose methods and actions designed to prevent the perpetrators from being identified and to make a pointed display of violence and the omnipresent repressive apparatus (p. 5).

- During the early eighties, a climate of terror spread across the country, characterized by extreme violence against communities and organized movements against which the people were completely defenseless. An atmosphere of constant danger totally disrupted the daily life of many families. Whether in the form of mass killings or the appearance of corpses bearing signs of torture, the horror was so massive and so flagrant that it defied the imagination (p. 9).

- Torture is described in conjunction with massacres and detentions. Besides seeking information, the purpose of torture is to destroy the victims' identities and either eliminate them or make them into accomplices to repression against their own neighbors and associates. In Guatemala, the social aspects of the use of torture represent an assault on the collective identity. In rural areas, torture sessions frequently were held in public, in front of relatives and neighbors, as a form of exemplary terror (p. 151).

- A central feature of the counterinsurgency strategy has been to place the blame and the responsibility on victims and survivors. To this end, the army's key tactics were propaganda and psychological warfare; structures such as civil patrols to militarize society and encourage conformity; and religious sects (p. 22).

• Threatening and torturing children was also a means of torturing their families. The torture of children was a means of forcing people to collaborate, inducing people to denounce others, and destroying community. It was a form of exemplary terror for their families and was an extreme demonstration of contempt for people's lives and dignity (p. 32).

• Death squads appeared in 1966 as part of the army's first massive counteroffensive against guerrilla forces. They were conceived as the operational branch of intelligence and served primarily to threaten, torture, and execute political opposition figures. One of the main impacts was to spread psychological terror among the population (p. 110).

• Community breakdown and displacement made it very difficult to continue to practice religious rituals and ceremonies. Fear of professing the Catholic faith, which the army considered subversive doctrine, was the most common obstacle to religious expression in rural areas (p. 46).

The abusive tactics at the heart of the "maniacal windstorm" in Guatemala are strikingly similar to those described earlier: from CIA manuals recommending assassinations, complete with big knives to slit throats; to admonitions that the only way to win the war is "to kill, kidnap, rob, and torture"; from SOA training manuals recommending that counterintelligence agents target family members as leverage over potential informants; to use of death squads, and the targeting of progressive religious as subversives worthy of extermination. An editorial in the *Boston Globe* titled, "Lesson in Terror," described the SOA training manuals:

Murder, extortion, torture — those are some of the lessons the US Army taught Latin American officers at the notorious School of the Americas in Columbus, GA. Recent revelations that the Pentagon trained police and military leaders in committing blatant atrocities describe a program that is beyond redemption.[9]

The REMHI report testifies to the central role played by SOA graduates in Guatemala in implementing these tactics and strategies of terror. For example, the creator of the infamous civil patrols is SOA graduate and U.S.-supported dictator Gen. Lucas García.

> The Guatemalan army formed the Civilian Self-defense Patrols in late 1981 as part of the counterinsurgency strategy. The main function of the PACs was to involve the communities in the army's anti-guerrilla offensive. The army realized that the insurgency enjoyed significant support from the civilian population. It therefore intended to use civil patrols to seal off communities from potential guerrilla penetration as well as remove the guerrillas from areas where they had already established a presence. The patrols began to operate under the government of Gen. Romeo Lucas García.[10]

Lucas García's rule corresponded to one of the most repressive periods in Guatemala's tortured history. The civil patrols, according to the report, were responsible for some of the most brutal violations of the war. Civil patrols participated in murders, torture, and other cruel treatment, forced disappearance, irregular detention, and threats. "Civil patrols, together with military commissioners, are implicated in one in five cases of deaths resulting from the persecution of people seeking refuge in uninhabited areas," the report states. "The civil patrols are identified as the perpetrators in nearly one in five massacres.... Taken together, these irregular government forces were responsible for one out of every four collective murders."[11]

"The civil patrols murdered many people in their own communities," the report says. The "killings were indiscriminate; anyone considered suspicious was liable to be targeted. These incidents were characterized by a disproportionate use of force, which was used against victims who were completely defenseless and who were often killed in front of their families."[12] "The mere presence of civil patrols as permanent entities in many communities," the report states, "had its effect on the children. From fear of aggression or death to the *normalization* of violence as a way of life, children were influenced by the warlike socialization patterns in a militarized environment."[13]

SOA connections to torture and terror through its graduates extend far beyond the Lucas García dictatorship and the repressive civil patrols he created. *Guatemala: Never Again!* documents the gruesome role played by Guatemalan intelligence agencies:

> The Guatemalan intelligence services played a pivotal role in the evolution of counterinsurgency policy. They comprise a complex network of military and police corps that permeated the social fabric (through agents, informers, and so forth), maintained their own hierarchies, and almost always enjoyed total autonomy of action. Military intelligence has played a key role in directing military operations, massacres, extrajudicial executions, forced disappearances, and torture. Throughout the armed conflict, intelligence officers and specialists were deeply involved in systematic human rights violations.[14]

The REMHI report also notes:

> The training curriculum...for intelligence agents included techniques for carrying out clandestine operations and abductions. Intelligence agents specialized in the logistics of abductions, the division of labor among the different members of the group; and coordinating rapid, clandestine strikes.[15]

The most notorious of the intelligence services was a division of the National Defense Staff known as the Army Intelligence Directorate or D-2, also called La 2.[16] "Its key agents," the report states, "were assigned to command positions in the army, giving military intelligence control over a wide range of material, technical, and human resources to carry out its own operations."[17] According to REMHI:

> La 2 [D-2] figures prominently in the worst incidents of violence; its dossier is replete with disappearances, murders, abductions, and torture. It conducted extensive espionage and

information-gathering operations by tapping telephones and operating a sophisticated computer network containing files on people, complete with their photographs, and information on their political and organizational affiliations.[18]

SOA graduates serving in influential posts are part of a consistent pattern linking the school to human rights abuses. Three top leaders and many officials of the fearsome D-2 are SOA graduates. SOA graduates featured in the report include three D-2 directors: Francisco Ortea Menaldo, César Augusto Cabrera Mejía, and Manuel Callejas y Callejas. Others in leadership positions include: Federico Sobalvarro Meza, César Quinteros Alvarado, Luis Felipe Caballeros Meza, Harry Ponce, Francisco Edgar Domínguez López, Eduardo Ochoa Barrios, Domingo Velásquez Axpuac, and José Manuel Rivas Ríos.

As was the case in El Salvador, SOA graduates are frequently at the center of high-visibility atrocities in Guatemala. Two of the three people named in the report as the intellectual authors of the 1991 murder of the internationally respected Guatemalan anthropologist Myrna Mack are SOA graduates. SOA graduate Marco Tulio Espinoza is held responsible for the disappearance of a guerrilla leader that nearly derailed the Guatemalan peace process. After the 1990 assassination of U.S. innkeeper Michael DeVine by Guatemalan military forces, President Serrano ordered charges brought against those responsible. He soon was frustrated, however, by stonewalling from the army and the D-2 and he blamed SOA graduate César Augusto Cabrera Mejía for blocking the investigation by denying records. SOA graduate Luis Miranda Trejo commanded the military base from which Captain Hugo Contreras, implicated in DeVine's murder, "escaped" from a high-security area.

Another high-visibility case is that of the 1992 torture and murder of Efraín Bamaca, husband of U.S. citizen Jennifer Harbury. According to testimony in the report, two SOA graduates, Ismael Segura Abularcach and Col. Ruano del Cid commanded the special forces that forced Bamaca, while a prisoner, to guide army patrols in search of guerrilla arms caches.

SOA graduate and CIA paid asset Col. Julio Roberto Alpirez was present during torture sessions and was linked to the killings of both Bamaca and Michael DeVine. Col. Alpirez, while on the CIA payroll, spent 1989 at the School of the Americas, and then returned in 1990 to Guatemala, where he continued working for the agency. It is likely that one important function of the SOA and SOA/WHISC is to recruit CIA assets from among the pool of Latin American officers attending the school. Panamanian dictator Manuel Noriega, SOA graduate and long-time CIA operative, would be another case in point. Returning to Alpirez, a *Washington Post* editorial, "Our Man in Guatemala," offered this summary of Alpirez's ties to the CIA:

> The CIA, learning of the atrocities [the murders of DeVine and Bamaca], contained and covered up the relevant information, ostensibly to protect "sources and methods." Officials at the State Department and National Security Council kept the story from the American wife. Word finally got out only as a result of disclosures by Rep. Robert Torricelli. It defies credulity that, at this late date in the United States Central American involvement, the CIA could still be recruiting killers of the sort that have made Guatemala's the region's bloodiest army. To hire an informant is one thing. To condone his criminality, by doing nothing to bring him to justice after two murders, is to lend official American approval — on the level where it counts most — to the Guatemalan military's criminal habit.[19]

The editorial also notes that the two murdered Americans "are two among 150,000 or more in a dirty war that the United States helped along mightily by conspiring to oust the elected leftist leader in 1954."[20]

I find this editorial particularly interesting for three reasons. First, it acknowledges U.S. involvement in a dirty, destructive war and the longevity of that involvement. Second, it offers evidence of a pattern of containment and cover-up by U.S. officials in response to revelations of atrocities, a dynamic that has been ever-present in the responses of Pentagon, White House, and SOA officials to documentation of SOA abuses. Third, the editorial suggests strongly that it was acceptable to "be recruiting killers" like Alpirez in the past as part of the Cold War mission but that such tactics are no longer necessary today. I will explore this perception and its relevance for the SOA/WHISC mission, past and present, in later chapters.

The character and conduct of Colonel Alpirez were defended by former Guatemalan Defense Minister and SOA graduate, General Héctor Gramajo. Gramajo described Alpirez as "a soldier above all." "He is the kind of officer you would want under your command."[21] Gramajo, while studying on a USAID grant at Harvard's Kennedy School of Government, was himself sued by Sister Diana Ortiz and a group of indigenous massacre survivors under the Alien Torts Act, which allows U.S. citizens and others residing in the U.S. to prosecute human rights abusers who are in this country for human rights violations. Since he never bothered to show up in court to defend himself he was found guilty by default and was ordered by a U.S. court to pay $47.5 million in damages (which he has never been forced to pay). It was Gramajo who drew up the plan for massacres in the highlands in 1982–83 and was responsible for seeing that these plans were carried out. The case of Diana Ortiz involves a U.S. nun who was abducted on November 2, 1989, raped and tortured by Guatemalan security forces, apparently *under the direction of a U.S. advisor*. Sister Ortiz's "crime" was that she taught Mayan children to read, write, and reflect on the Bible in the context of their Mayan heritage.

Sister Ortiz worked tirelessly, including a lengthy fast, to force the United States government to reveal what they knew about her case. What she discovered and experienced along the way was a pattern of trickery, secrecy, and cover-ups. The U.S. Embassy in Guatemala "initiated a smear campaign against Ortiz immediately upon report of her abduction."[22] One document, dated March 19, 1990, urgently expressed the need to "close the loop on the issue of the North American named by Ortiz ... THE EMBASSY IS VERY SENSITIVE ON THIS ISSUE." Two completely blacked-out pages follow.[23]

General Gramajo, who was Minister of Defense and head of the armed forces, blamed Ortiz's hundred-plus burn marks on a failed lesbian affair. Gramajo's role in committing or justifying human rights atrocities apparently fits well within the framework of the SOA's Cold War mission.[24] Two years after his involvement in covering-up Sister Diana Ortiz's abduction, torture, and rape, the School of the Americas sent a powerful anti-human rights message by inviting Gramajo to deliver the commencement address to graduate officers for the Command and General Staff College at the school. *The Bayonet,* Fort Benning's authorized newspaper, reported:

> Following the invocation, the guest speaker, retired Gen. Héctor Gramajo from Guatemala, addressed the audience of graduate officers. Gramajo voiced concern for a continued vigilance in the Americas against communism and drug trafficking. Comparing the current state of communism to a dragon, Gramajo said the crumbling of the Berlin Wall signaled the beheading of the dragon; however, its tail is still poised to deliver a devastating blow to the countries of Latin America.[25]

Guatemalan Bishop Juan Gerardi was murdered in April 1998, two days after he unveiled the REMHI report from which I have quoted extensively. SOA graduate Byron Lima Estrada was one of four military officers convicted of his murder on June 8, 2001. "We are collecting the people's memories," Bishop Gerardi said on the occasion of the release of the report, "because we want to contribute to the construction of a different country. This path," he noted prophetically, "was and continues to be full of risks, but the construction of the Kingdom of God entails risks, and only those who have the strength to confront those risks can be its builders."[26] His words are particularly relevant to U.S. citizens who need to hold the SOA and U.S. foreign policy planners accountable for their role in numerous atrocities:

> To open ourselves to truth and to face our personal and collective reality are not options that can be accepted or rejected. They are indispensable requirements for all people and societies that seek to humanize themselves and to be free. They make us face our most essential human condition: that we are

sons and daughters of God, called to participate in our Father's freedom. Years of terror and death have displaced and reduced the majority of Guatemalans to fear and silence. *Truth* is the primary word, the serious and mature action that makes it possible for us to break this cycle of death and violence and to open ourselves to a future of hope and light for all.[27]

Chapter 5

More Evidence
and Key Questions

I want to offer four additional examples that demonstrate that the disturbing links between U.S. foreign policy, the School of the Americas/WHISC, SOA graduates, and repression and terror extend well beyond El Salvador and Guatemala. It is not my intent to provide an exhaustive country-by-country list of SOA graduates involved in human rights abuses but rather to focus on the role of the school in several countries as a way of highlighting its importance to U.S. foreign policy. The collective weight of these examples should be sufficient to dispel several myths propagated by SOA/WHISC defenders: the "bad apple theory" that says that the school should not be blamed for the unfortunate actions of a tiny minority of its graduates; the charge that critics of the SOA cite only abuses from the distant past; and, the public relations assertion that a primary mission of the SOA is defense of democracy and human rights.

Death Squads and Cover-ups

Honduras played a key role in U.S. foreign policy in the 1980s and was central to CIA director William Casey's efforts to "waste" Nicaragua by subjecting its people to attacks from the U.S.-created, armed, funded, and trained contras. Honduran territory was the safe-haven from which the largest contra group waged a terrorist war against the people of Nicaragua with CIA training manuals and Argentina's generals serving as guides. As part of its effort to destabilize Nicaragua, the U.S. militarized Honduras. In doing so the United

States supported a series of brutal military-dominated governments and the CIA, with SOA graduates playing important roles, set up repressive internal mechanisms, including death squads.

A detailed investigation by Gary Cohn and Ginger Thompson of the *Baltimore Sun* confirms allegations that the United States was intimately connected to death squads and torturers in Honduras. According to their report, the "CIA was instrumental in training and equipping Battalion 316," a secret army unit that was home to Honduran death squads:

> The intelligence unit, known as Battalion 316, used shock and suffocation devices in interrogations. Prisoners often were kept naked and, when no longer useful, killed and buried in unmarked graves. Newly declassified documents and other sources show that the CIA and the U.S. Embassy knew of numerous crimes, including murder and torture, yet continued to support Battalion 316 and collaborate with its leaders.[1]

At least nineteen of the ranking Honduran officers linked to death squad Battalion 316 are SOA graduates, including battalion founder General Luis Alonso Discua. José Valle, a School of the Americas graduate, a member of Battalion 316, and an admitted torturer, told Robert Richter that he took "a course in intelligence at the School of the Americas" in which he saw "a lot of videos which showed the type of interrogation and torture used in Vietnam.... Although many people refuse to accept it," Valle said, "all this is organized by the U.S. government."[2]

The disturbing claim that U.S. personnel supervised or taught torture techniques was evident in Guatemala in the case of Sister Diana Ortiz discussed earlier. In the midst of her torture she reports the arrival of a man called Alejandro, who spoke Spanish with a U.S. accent, and to whom her torturers deferred when he told them they had picked up the wrong person. The claim is also strengthened by the testimony of an SOA graduate who was interviewed as part of a documentary *Inside the School of Assassins*. He spoke on film on the condition that his face and name not be used:

The school was always a front for other special operations, covert operations. They would bring people from the streets [of Panama City] into the base and the experts would train us on how to obtain information through torture. We were trained to torture human beings. They had a medical physician, a U.S. medical physician which I remember very well, who was dressed in green fatigues, who would teach the students . . . [about] the nerve endings of the body. He would show them where to torture, where and where not, where you wouldn't kill the individual.[3]

Another troubling feature of this despicable tale is that the U.S. ambassador to Honduras during this period, John Negroponte, was fully aware of these atrocities, supported them, and tried to cover them up. "Time and time again during his tour of duty in Honduras from 1981 to 1985," the *Baltimore Sun* series notes, "Negroponte was confronted with evidence that a Honduran army intelligence unit, trained by the CIA, was stalking, kidnapping, torturing, and killing suspected subversives."[4] The article continues:

Rick Chidester, then a junior political officer in the U.S. Embassy in Tegucigalpa . . . compiled substantial evidence of abuses by the Honduran military in 1982, but was ordered to delete most of it from the annual human rights report prepared for the State Department to deliver to Congress. Those reports consistently misled Congress and the public. "There are no political prisoners in Honduras," the State Department asserted falsely in its 1983 human rights report. The reports to Congress were carefully crafted to convey the impression that the Honduran government and military were committed to democratic ideals. It was important not to confront Congress with evidence that the military was trampling on civil liberties and murdering dissidents. The truth could have triggered congressional action under the Foreign Assistance Act, which generally prohibits military aid to any government that "engages in a consistent pattern of gross violations of internationally recognized human rights."[5]

Allegations that Negroponte suppressed information about human rights atrocities were confirmed in a 1997 CIA Inspector General's report.[6] A source cited in the report explained the deception, saying that reporting corruption, murders, and executions would "reflect negatively on Honduras and not be beneficial in carrying out U.S. policy."[7] It is rather sobering to realize, and we should keep such things in mind when we evaluate the significance of the SOA's recent name change, that John Negroponte is George W. Bush's nominee to serve as U.S. ambassador to the United Nations. No confession. No repentance. No accountability. No restitution. A promotion. Linked to the WHISC we might say: No confession. No repentance. No accountability. No restitution. Same location. Similar curriculum. New name.

Chile reveals a similar pattern, including human rights abuses, a central SOA role, involvement by other high-level U.S. officials, and rewards offered to guilty parties. In his two-part exposé in *Harper's Magazine* on Henry Kissinger ("The Making of a War Criminal" and "Crimes against Humanity"), Christopher Hitchens includes numerous details on the U.S. role in the 1973 coup that overthrew the democratically elected leader of Chile and implanted a brutal military dictatorship. Kissinger had stated his personal disdain for Chilean democracy after the election of socialist politician Salvador Allende. He saw no reason, he said, why a country should be allowed to "go Communist due to the irresponsibility of its own people."[8]

Hitchens documents that the CIA working under orders of President Nixon and with personal guidance from Kissinger arranged the coup to overthrow Salvador Allende, the democratically elected president of Chile, and the murder of General René Schneider, the head of the Chilean Armed Forces, who refused to carry out the coup in deference to the Chilean constitution. According to Hitchens:

> A series of Washington meetings, within eleven days of Allende's electoral victory, essentially settled the fate of Chilean democracy. After discussions with Kendall [the president of Pepsi-Cola], with David Rockefeller of Chase Manhattan, and with CIA director Richard Helms, Kissinger went with Helms

to the Oval Office. Helms's notes of the meeting show that Nixon wasted little breath in making his wishes known. Allende was not to assume office. "Not concerned risks involved. No involvement of embassy. $10,000,000 available, more if necessary. Full-time job — best men we have. . . . Make the economy scream. 48 hours for plan of action."

He continues:

> Declassified documents show that Kissinger . . . took seriously this chance to impress his boss. A group was set up in Langley, Virginia, with the express purpose of running a "two-track" policy for Chile, one the ostensible diplomatic one and the other . . . a strategy of destabilization, kidnapping, and assassination designed to provoke a military coup. There were long- and short-term obstacles. . . . The long-term obstacle was the tradition of military abstention from politics in Chile, a tradition that marked off the country from its neighbors. Such a culture was not to be degraded overnight. The short-term obstacle lay in the person of one man: General René Schneider. As chief of the Chilean Army, he was adamantly opposed to any military meddling in the electoral process. Accordingly, it was decided at a meeting on September 18, 1970, that General Schneider had to go.[9]

Eight days before Allende's scheduled confirmation as president the "track two" group received a CIA cable from Santiago saying:

> It is firm and continuing policy that Allende be overthrown by a coup. It would be much preferable to have this transpire prior to 24 October but efforts in this regard will continue vigorously beyond this date. We are to continue to generate maximum pressure toward this end utilizing every appropriate resource. It is imperative that these actions be implemented clandestinely so that the USG [United States Government] and American hand be well hidden.[10]

President Allende and General Schneider were both murdered as part of the successful coup, but not before Allende took office. Following the coup agents of the new repressive government headed by General Augusto Pinochet killed or disappeared thousands of workers, students, and progressive religious. It also tracked and assassinated exiled Chilean citizens throughout Latin America and in the United States as part of Operation Condor. "Condor," Hitchens notes, "was a machinery of cross-border assassination, abduction, torture, and intimidation coordinated among the secret police forces of Pinochet's Chile, Alfredo Stroessner's Paraguay, Jorge Rafael Videla's Argentina, and other regional caudillos." This "internationalization of the death squad principle," Hitchens notes, resulted in the murder of numerous dissidents. *"United States government complicity has been uncovered at every level of this network."*[11] Kissinger, Hitchens says, "approved the internationalization of the death-squad principle,"[12] and should be tried as a war criminal for his role in Chile as well as numerous other atrocities carried out under his direction in Vietnam, East Timor, Indonesia, and elsewhere. Once again, however, as was the case with Negroponte, the U.S. rewards possible war criminals instead of holding them and the foreign policy they service accountable. Kissinger's "advice is sought, at $30,000 an appearance, by audiences of businessmen, and academics and policymakers."[13]

The whole story told by Hitchens in all its gruesome detail is necessary reading but can't be recounted here. There is, however, an additional piece of the story that Hitchens did not tell. General Pinochet was not only put in power courtesy of a U.S. coup, he staffed his repressive regime with numerous SOA graduates. He was arrested and indicted on human rights grounds in 1998 on the orders of a Spanish judge. The Spanish lawyers who presented the charges against Pinochet also requested the indictment of thirty other high-ranking officials of the Chilean dictatorship for crimes of genocide, terrorism, torture, and illegal arrest followed by disappearance. Ten of those to be indicted are graduates of the School of the Americas. Although Pinochet is not himself an SOA graduate, the Chilean dictator is held in high esteem at the school. In 1991, visitors could view a note from Pinochet and a ceremonial sword donated by him on display in the office of the Commandant.

Another El Salvador

Recently painted graffiti on a wall in Bogotá, Colombia, shows a flower with the inscription: "Plan Colombia, plan de muerte, no asesinaran la esperanza," which means "Plan Colombia, plan of death, don't kill the hope." The country with the dubious distinction of having sent the most students to the SOA (more than 10,000) over the past fifty-five years is Colombia, one of the most violent and corrupt countries in the world.

The United States has a long history of involvement in Colombia's violence, including support for brutal counterinsurgency campaigns justified in the name of fighting drugs. Javier Giraldo, S.J., director of a Colombian human rights organization, in his book *Colombia: The Genocidal Democracy*, writes about "the false conclusion that violence in Colombia is linked to drug traffic." Most of the violence, he notes, is linked "to paramilitary groups which operate as auxiliaries to the army and police."[14] Noam Chomsky, in the Introduction to Giraldo's book, details the real war being fought under the cover of the drug war:

> In July 1989, the U.S. State Department announced plans for subsidized sales of military equipment to Colombia, allegedly "for antinarcotics purposes." The sales were "justified" by the fact that "Colombia has a democratic form of government and does not exhibit a consistent pattern of gross violations of internationally recognized human rights." A few months before, the Commission of Justice and Peace that Father Giraldo heads had published a report documenting atrocities in the first part of 1988, including over 3,000 politically motivated killings, 273 in "social cleansing" campaigns. Political killings averaged eight a day with seven people murdered in the homes or in the street and one "disappeared." Citing this report, the Washington Office on Latin America (WOLA) added that "the vast majority of those who have disappeared in recent years are grassroots organizers, peasant or union leaders, leftist politicians, human rights workers, and other activists," over 1,500 by the time of the State Department's praise for Colombia's democracy and

its respect for human rights. During the 1988 electoral campaigns, 19 of 87 mayoral candidates of the sole independent political party, the UP, were assassinated, along with over 100 of its other candidates. The Central Organization of Workers, a coalition of trade unions formed in 1986, had by then lost over 230 members, most of them found dead after brutal torture. But the "democratic form of government" emerged without stain, and with no "consistent pattern of gross violations" of human rights.[15]

This pattern continues. Despite one of the worst human rights records of any nation and in the context of ongoing collaboration between Colombian military and paramilitary forces implicated in horrific violence, the U.S. Congress approved a $1.3 billion aid package ("Plan Colombia") in 2000.

Plan Colombia, like earlier aid packages, is promoted by the Pentagon under the guise of fighting the drug war. The principal victims, however, continue to be those targeted in a brutal counterinsurgency war rooted in issues of poverty, land rights, and oil. The majority of the funds from Plan Colombia go to U.S. weapons corporations such as United Technologies and Sikorsky for the sale of Huey and Black Hawk helicopters, and to Monsanto, which produces the herbicide "glyphosate" used in fumigations. Although the U.S. government says fumigations target coca crops (the raw material for cocaine), hundreds of thousands of acres of food crops and rainforests are being decimated.

Massacres and human rights abuses committed by paramilitaries working in conjunction with the Colombian military have been a staple feature of the war in Colombia for decades. Giraldo writes:

> As the [paramilitary] project advanced through the countryside, rural communities were told they had three options: join the paramilitaries, leave the region or die. It soon became clear that the paramilitary project enjoyed support at the highest echelons of government. Paramilitary bases were constructed next to military bases, meetings with campesinos were called by soldiers and run by paramilitaries or vice versa, census data and

lists of campesino families and . . . [property] owners elaborated by the army turned up in the possession of the paramilitaries, and individuals detained by soldiers were turned over to the paramilitaries.[16]

The great advantage of the paramilitary strategy is that it provides cover for repressive governments and the political and economic interests they represent. "The government was able to successfully conceal its role in and evade responsibility for crimes," Giraldo notes, "by entrusting much of the 'dirty work' to armed civilian groups which began to operate under the clandestine coordination of the army and police."[17] "In the past few years the Colombian military has gotten out of directly waging the dirty war, and at the same time there has been a commensurate rise in the size and ferocity of the paramilitaries," according to Andrew Miller of Amnesty International. "And it is amply documented that even if independently financed, the paramilitaries work hand in hand with the government forces."[18]

Massacres and human rights abuses committed by paramilitaries working in conjunction with the Colombian military have doubled since 1998. According to Human Rights Watch's report "The Ties That Bind: Colombia and Military-Paramilitary Links":

Far from moving decisively to sever ties to paramilitaries, Human Rights Watch's evidence strongly suggests that Colombia's military high command has yet to take the necessary steps to accomplish this goal. Human Rights Watch's information implicates Colombian Army brigades operating in the country's three largest cities, including the capital, Bogotá. If Colombia's leaders cannot or will not halt these units' support for paramilitary groups, the government's resolve to end human rights abuse in units that receive U.S. security assistance must be seriously questioned. . . . Together, evidence collected so far by Human Rights Watch links half of Colombia's eighteen brigade-level army units (excluding military schools) to paramilitary activity. These units operate in all of Colombia's five divisions. In other words, military support for paramilitary activity remains national in scope and includes areas where units receiving or scheduled to receive U.S. military aid operate.[19]

SOA graduates are featured centrally in both the massacres and the drug war in Colombia. Here is a sampling of evidence linking SOA graduates to atrocities in Colombia:[20]

- General Mario Montoya Uribe, a graduate and former instructor at the SOA with a history of ties to paramilitary violence, commands the Joint Task Force South that includes the infamous 24th Brigade. The 24th Brigade is ineligible for U.S. military aid due to its complicity in paramilitary violence. A leading Colombian newspaper identifies General Montoya as "the military official responsible for Plan Colombia."[21] He is responsible for all military activities in the Putumayo region, the focal point of Plan Colombia. Putumayo is important because it is where Colombia, Venezuela, and Ecuador meet. Colombian Ricardo Esquiva, director of Justa Paz, told a Witness for Peace delegation: "That is why the fighting is so intense in Putumayo. Whoever controls the Putumayo area will have control of South America."[22]

- More than 100 of the 246 Colombian officers cited for war crimes by an international human rights tribunal in 1993 are SOA graduates.

- SOA graduate Gen. Jaime Ernesto Canal Alban, commander of the Third Brigade, is associated with numerous massacres. "The Ties That Bind" report released in February 2000 named at least seven SOA graduates for involvement in paramilitary groups. Canal was involved in helping to establish a paramilitary group known as the "Calima Front." Canal's brigade supplied the Front with weapons and intelligence and carried out joint operations. In 1999, the Calima Front seized and executed community leader Noralba Gaviria Piedrahita. The following month, authorities discovered the mutilated and dismembered bodies of seven men near Tulu, also killed by members of the Calima Front. The Front has been responsible for 2,000 forced disappearances and at least forty executions since 1999. In addition to his involvement with the Calima Front, Canal was in command of soldiers who entered a home and killed five civilians during a birthday party of a fifteen-year-old child in 1998.

- The Human Rights Watch report cites SOA graduate Gen. Carlos Ospina Ovalle, former commander of the 4th Brigade, for "extensive evidence of pervasive ties" to paramilitary groups involved in human rights abuses throughout 1999. Ospina was the commander of the 4th Brigade when troops massacred at least eleven people and burned 47 homes in El Aro. Paramilitary forces returned to El Aro and kidnapped Aurelio Areiza. They gouged out his eyes and cut off his tongue and testicles before killing him. Ospina is in charge of "security assistance" allocated by the United States.

- The Human Rights Watch Report also cites Major Alvaro Cortés Morillo and Major Jesús María Claviho, both SOA graduates, for ties to paramilitary groups in 1999 through extensive cell phone and beeper communications as well as regular meetings on military bases.

- SOA graduate Col. Jorge Plazas Acevedo, former chief of intelligence for the Colombian Military's 13th Brigade, is being tried by the Prosecutor General of Colombia for the 1998 kidnapping and murder of Jewish business leader Benjamin Khoudari.

- According to the 2000 State Department Report on Human Rights in Colombia, SOA graduates Major David Hernández Rojas and Captain Diego Fino Rodríguez are being prosecuted in civilian courts for the March 1999 murders of Antioquia peace commissioner Alex Lopera and two others. Both are members of the Colombian Military's 4th Brigade known to have extensive ties to paramilitary groups.

- In February 2001, SOA graduate Hernán Orozco was sent to prison by a military tribunal for complicity in the Maripan torture and massacre of 30 peasants by a paramilitary group.

Critics of Plan Colombia fear that Colombia *may* become the next Vietnam. Although this is a legitimate concern it seems more accurate to say that Colombia *is* the new El Salvador. Adam Isacson of the Center for International Policy compiled the following list of comparisons:

- El Salvador is the size of Massachusetts and in the 1980s had less than 5 million people. Colombia is 53 times larger than El Salvador (the size of Texas, New Mexico, and Oklahoma combined) and has 40 million people.

- El Salvador's economy and politics are dominated by a small landholding elite. Colombia's economy and politics are also dominated by a small elite. The wealthiest 10 percent earn 47 times more money each year than the poorest 10 percent. (The figure in the United States is 16, Canada is 8, most of Western Europe is less than 7.) 1.3 percent of the landholders control 50 percent of the land.

- El Salvador's military ruled the country directly from the 1930s to the 1980s, and committed thousands of human rights abuses during the 1980s. Colombia's military has only ruled the country once, for three years in the 1950s. Civilian leaders have very little control over the armed forces, however. Colombia's military committed thousands of human rights abuses in the 1990s, but its share of direct abuses dropped sharply in the late 1990s. This drop went hand-in-hand with a sharp rise by right-wing paramilitary groups.

- El Salvador's death squads, tied to the wealthy elite and the military, operated clandestinely and committed innumerable human rights abuses. Colombia's paramilitary groups, tied to the wealthy elite and the military, operate out in the open like a guerrilla army. They are now responsible for 85 percent of killings associated with Colombia's conflict.

- Instead of offering economic aid to alleviate the causes of El Salvador's conflict, the Reagan administration chose to fight Communism by strengthening the Salvadoran military. Instead of offering economic aid (and meeting domestic needs for drug treatment) to address the reasons why peasants in Colombia grow coca, the Clinton and Bush administrations have chosen to fight drug abuse by strengthening the Colombian military.

- At its peak in 1984, U.S. aid to El Salvador's military was $1 million per day. At its beginning in 2000–2001, U.S. aid to Colombia is nearly $2 million per day.

- The U.S. military presence in El Salvador was limited to a maximum of 55 advisors. These troops could offer advice in combat situations. The U.S. military presence in Colombia is limited to a maximum of 500 uniformed personnel and 300 contractors. These troops must avoid combat situations and the United States is careful to avoid any risk of casualties. They are serving mainly as trainers and intelligence-gatherers. The contractors, civilians who work for U.S.-based companies, work as spray-plane and helicopter pilots, logistics personnel, and intelligence gatherers, among other duties.

- During the 1980s, El Salvador was the number one "feeder school" for the School of the Americas. Helicopter courses were a big source because the U.S. had provided the Salvadorans with many Vietnam-era helicopters. Colombia is the number one "feeder school" for the School of the Americas (now called the Western Hemisphere Institute for Security Cooperation) today. However, most of the more than 4,000 Colombians the United States trains each year are taught by U.S. Special Forces who offer courses on Colombian soil. Again, helicopter courses are a big source of training because the United States provides Colombia with many Vietnam-era helicopters, plus a fleet of brand new Black Hawk helicopters, valued at $15 million apiece.

- The Reagan administration was hostile to the Contadora and Arias peace processes, which sought to end El Salvador's war through negotiations. The Clinton and Bush administrations have offered only lukewarm support for Colombian President Andrés Pastrana's efforts to negotiate peace with guerrilla groups. Some officials openly criticize the peace process.

- The Reagan administration issued patently false certifications to Congress claiming that the Salvadoran military's human rights record was improving. Instead of offering a similar certification, the Clinton administration waived the human rights conditions. (Congress included a waiver clause in the 2000 aid package law.)[23]

Repression in Colombia has always been linked to issues of land and resources. Colombian human rights activist Héctor Mondragón, himself a torture victim at the hands of an SOA graduate, said recently that "in the last 15 years the largest landholders in Colombia have grown from holding 32% of the land to holding 45%." Land concentration in the context of Colombia's present crisis has less to do with production and more to do with proximity to "oil reserves and wells," "mega projects," and "foreign investment." For example, "many of the recent massacres . . . are related to their closeness to the oil reserves" where the companies "have strong ties with the paramilitary groups." In these areas "people were massacred by the paramilitaries." "And there was not any combat, there was no battle. And meanwhile the Navy and the Army had the region surrounded. Really the only thing they did was guarantee that the massacre could happen, they secured the area so that the massacre could happen." "When the US supports the Colombian state they are supporting these kinds of genocides" under the "pretext" of "the war against drugs."[24]

That the drug war is a "pretext" is evident in a 1999 General Accounting Office (GAO) report: "despite two years of extensive herbicide spraying, U.S. estimates show [that] . . . net coca cultivation actually increased 50%."[25] A *New York Times* editorial notes that Plan Colombia "is skewed toward using military force to shut down the drug trade in Colombia, an approach that could entangle American troops in that nation's protracted civil war while doing little to stem the flow of drugs." "The bulk of the federal government's $19.2 billion annual drug-fighting budget," according to the editorial, "is still spent on interdiction and enforcement. Yet the number of hard-core users of cocaine has remained steady over the last decade...."[26]

Why escalate a drug war if the drug war is failing? Not surprisingly, U.S. business groups connected to weapons and oil benefit from Plan Colombia. Winifred Tate, in an article "Repeating Past Mistakes: Aiding Counterinsurgency in Colombia," writes:

> The heavy hand of corporate money is also clearly visible in the Colombia debate, particularly in the case of helicopter and oil companies. At stake are nearly $400 million in contracts for helicopters such as the Blackhawk. Oil companies, meanwhile, claim to have lost hundreds of millions of dollars' worth of oil to guerrilla sabotage; in addition, the firms are interested in new drilling in conflictive areas. During the House hearings on Colombia, Occidental Petroleum Vice President Lawrence Meriage was one of very few nongovernmental witnesses to testify before Congress. As a leader of the U.S.-Colombia Business Partnership, founded in 1996 to represent U.S. companies with business interests in Colombia, Meriage led business sector support for the package. During the Senate debate, minimal time was devoted to considering the wisdom of massive military assistance....[27]

Marc Cooper writes about the extensive U.S. military and intelligence presence in Colombia:

> Walk into the marble-floored and track-lit headquarters of Colombia's national antinarcotics police and the generosity of ... [U.S.] aid, as well as the incestuous relationship between Washington and Colombia's military machine are suddenly evident. Outside the door of Commanding Gen. Gustavo Socha's office, mounted on a tripod, is an oversize photo of a grinning George W. Bush celebrating his election. Next to it is a full-

color promotional illustration of a US-made Black Hawk attack helicopter. In the general's waiting room, visitors are attended to by a young, uniformed press officer, a polished graduate of the recently renamed School of the Americas, run by the US Army. Also present is an equally young security officer just returned from an intelligence training course at Lakeland Air Force Base in Texas.

In case there's any doubt about the level of American involvement here, the office adjoining General Socha's is occupied by a...veteran trainer at the School of the Americas, the ex-colonel now works with the State Department's Narcotics Affairs Section and is deployed as a full-time advisor to General Socha.

Countless other federal drug and intelligence agents also work in Colombia. In addition there are a couple of hundred or more US military advisors training three new elite battalions of the Colombian Army. Dozens of US choppers are also arriving here: one fleet of "Super Hueys," mostly for the Colombian Army, and a squadron of top-of-the line Black Hawks, allocated mostly to Socha's antidrug troops. Along with them come an unknown number of private contract US pilots and helicopter technical crews. Another batch of private contract Americans are here to fly the crop-dusters that spray toxic herbicides over the coca-rich countryside. Supporting this operation are four new so-called Forward Operating Locations — US military intelligence outposts — in Ecuador, Aruba, Curaçao and El Salvador.[28]

This brief look at Colombia demonstrates that human rights atrocities linked to SOA graduates are not only a past problem but a present reality. The strikingly similar tactics employed in El Salvador in the 1980s and in Colombia today tell us that the school and U.S. foreign policy generally are still committed to destructive counterinsurgency practices when they are deemed necessary. The Colombian people are paying a heavy price for U.S. arrogance and goals. As a U.S. embassy official in Colombia noted: "The U.S. and

Colombia have different priorities. Colombia has peace as a priority. We have narcotics."[29] Not only narcotics, but the needs of oil companies, helicopter producers, chemical companies, and a military in search of enemies explain U.S. militarization of Colombia.

A Preference for Dictators

A careful reader may have noticed that the word dictator has surfaced frequently within this and previous chapters in reference to the School of the Americas. Guatemala's dictator, General Lucas García, is an SOA graduate and founder of the civil patrol system responsible for numerous atrocities and other massacres. Former Panamanian dictator and CIA asset General Manuel Noriega was an SOA graduate, as was Argentina's General Leopoldo Galtieri. Galtieri headed Argentina's military junta during a wave of terror and his officers trained the Nicaraguan contras at the behest of CIA director William Casey. Pinochet, although not an SOA alumnus, surrounded himself with SOA graduates and was revered by the school's commandant. Bolivian dictator Hugo Banzer was another SOA graduate who, as I will demonstrate in the next chapter, worked with the CIA to set up an Operation-Condor-type organization to track, intimidate, and, if necessary, kill progressive religious workers.

Former Representative Joseph Kennedy is right when he says that the "U.S. Army School of the Americas...is a school that has run more dictators than any other school in the history of the world."[30] SOA supporters have often cited the large number of military heads of state with pride. Many of their portraits hung on an infamous "Hall of Fame" that lined the school's stairwell. A major justification for the school is that cultivating relationships with Latin American military officials allows the United States to exercise influence and promote vital security interests. What better confirmation that military-to-military ties between Latin American officers and U.S. counterparts bear fruit than to have SOA graduates appear as heads of state in country after country. As nasty revelations concerning the school and its graduates came to the surface, however, so too did contradictions. Speaking during a House debate on the SOA, Joseph Kennedy pointed to the flawed reasoning of SOA supporters:

> They boast about the fact that 10 separate heads of state throughout Latin America were graduates of the School of the Americas. Not one of them was elected through a democratic election, and in many cases they actually overthrew the civilian governments that brought them into power.[31]

The SOA and its graduates were at the heart of numerous human rights atrocities in our hemisphere for decades, and their role and similar tactics are employed today in Colombia. In light of evidence presented thus far, readers may be asking themselves questions like these: Why did U.S. foreign policy conspire with, guide, fund, train and provide ideological cover for brutal militaries and paramilitaries throughout Latin America? Is this still going on? Did the SOA change? Has it really closed? Will the Western Hemisphere Institute for Security Cooperation be different? What is at stake in Colombia? How have the end of the Cold War and other geopolitical changes impacted U.S. foreign policy in terms of strategy and tactics? What does this mean for the SOA/WHISC? It is to these questions that I now turn.

Chapter 6

Geopolitics and the SOA/WHISC

Foreign Policy Stage 1

U.S. foreign policy, including military policy, has changed significantly since the founding of the U.S. Army School of the Americas in 1946. The SOA/WHISC is best understood as an evolving institution and flexible instrument of U.S. foreign policy in Latin America. Its past and present role can be best understood in light of four stages in U.S. foreign policy. Each stage has been influenced by and reflected the interests of powerful corporations and each has resulted in enormous pain and suffering for poor people in many parts of the so-called third world. The school's role within each stage reflects shifting geopolitical needs and opportunities.

- Stage 1 (roughly 1946 to 1979) was a period of militarization and dictatorship in which stable investment climates were created or maintained through military violence.

- Stage 2 (roughly 1980–90) was marked by a two-track strategy in which the U.S. intensified military repression where needed (places like Central America) and utilized economic leverage and economic institutions such as the IMF and World Bank as foreign policy instruments to force nations to comply with the wishes of powerful economic actors.

- Stage 3 (roughly 1991 to 1997) was a period of downsizing for many militaries in Latin America as various forms of economic power and leverage became the principal and preferred instruments of U.S. foreign policy. The economic rules imposed through repressive militarization and structural adjustment policies in the 1980s bore fruit and were institutionalized in free trade agreements such as NAFTA.

- Stage 4 (roughly 1998 to the present) is marked by a two-track foreign policy based on further institutionalization of economic leverage through the WTO and other free trade agreements, and remilitarization. Remilitarization responds to problem areas such as Colombia, instability that results from corporate-led globalization, and the resurgent power of the U.S. military-industrial-congressional complex.

Within the framework of these four stages, and the geopolitical changes they reflect, we can make sense out of the complex world of U.S. foreign policy and the SOA/WHISC, including: the preference for dictators, the war against progressive religious, and the repressive tactics employed in Central America in the 1960s, '70s and '80s; the transition from military governments to restricted democracy, the flourishing of human rights rhetoric, and the increased and preferred reliance on economic rather than military actors as instruments of foreign policy in the age of globalization; and, contradictions and strains in the system that account for a resurgence of militarization today in Colombia and elsewhere. Stage 1 is the subject of the present chapter. Stages 2–4 will be discussed in chapter 7.

Mal-development and Military Dictatorship

Following World War II the U.S. government sought to develop and democratize Japan and Europe while militarizing the so-called third world in defense of economic and strategic interests. The Chair of Standard Oil warned in 1946, the year the SOA opened, that U.S. private enterprise had to "strike out and save its position all over the world, or sit by and witness its own funeral..." and to "assume the responsibility of the majority stockholder in this corporation known as the world." The goal of U.S. foreign policy was, he said, to insure the "safety and stability of our foreign investments."[1] The School of the Americas' original mission was to insure stability by building and solidifying relationships between the U.S. and Latin American militaries through military training. Two years after the school's opening George Kennan, the most important U.S. foreign policy planner in the post–World War II period, stated clearly that U.S. foreign policy had to be a blunt instrument if the United States were to maintain its privileged position in a world of striking inequalities:

> We have about 50% of the world's wealth, but only 6.3% of its population.... In this situation, we cannot fail to be the object of envy and resentment. Our real task in the coming period is to devise a pattern of relationships which will permit us to maintain this position of disparity without positive detriment to our national security. To do so we have to dispense with all sentimentality and day-dreaming; and our attention will have to be concentrated everywhere on our immediate national objectives. We need not deceive ourselves that we can afford today the luxury of altruism and world-benefaction.

Speaking specifically about Asia, but with relevance for Latin America, Kennan added:

> We should cease to talk about vague and ... unreal objectives such as human rights, the raising of living standards and democratization. The day is not far off when we are going to have to deal in straight power concepts. The less we are hampered by idealistic slogans, the better.[2]

Although preoccupied with rebuilding war-ravaged Europe and Japan, the United States had "vital interests" throughout the third world where social conditions such as hunger, poverty, and inequality fed deep desires for change. Movements formed among students, workers, and peasants. Some were influenced by communist ideas or groups but most responded to blatant injustices rooted in their colonial experience and in the failures of the capitalist or semi-capitalist systems under which they lived and died. In this context, the United States put its considerable ideological, logistical and financial resources behind repressive militaries in the name of stability and fighting communism. Protests were inevitable, however, as Kennan stated bluntly in 1948, and the situation grew worse as U.S. foreign policy set out to establish and defend systems that fostered inequalities and sought to impose stability at the expense of democracy, freedom, development, and human rights.

One graphic example of mal-development that led and leads to a repressive military response concerns agriculture. For many decades before NAFTA the United States put unrelenting pressure on third world countries to ignore domestic agricultural production in favor of imports from the United States and other development goals. The U.S. was dumping so much surplus grain around the world in the decade following World War II that by 1957 a State Department official warned the Senate:

> We have a real problem in connection with the dependency created by some of these [food aid] programs. If dependence upon the U.S. goes too fast...we may be helping weaken a country and its economy in the long run.

Senator Hubert Humphrey responded:

> I have heard this morning that people may become dependent on us for food. I know that was not supposed to be good news. To me, that is good news, because before people can do anything, they have got to eat. And if you are looking for a way to get people to lean on you and be dependent on you, in terms of their cooperation with you, it seems to me that food dependence would be terrific.[3]

The CIA shared Humphrey's enthusiasm for food dependency as an instrument of power. In the midst of a world food crisis in 1974 the agency wrote that world food shortages could give the United States "virtual life and death power over the fate of the multitudes of the needy." "Twenty years ago," the CIA noted, "North America exported mainly to Western Europe; most other regions were basically self-sufficient. Now the whole world has become dependent on North America for grain...."[4]

Colombia offers a dramatic example of the relationship between policies that undermine indigenous agriculture and military repression. In Colombia during Stage 1 U.S. food aid programs encouraged Colombian marketers to sell wheat at prices low enough to eliminate the greater part of Colombian domestic production. Between 1955 and 1971 Colombia imported 1,140,000 tons of wheat. At the same time, prices received by Colombian farmers declined sharply. While in 1954 Colombia produced more than 160,000 tons for internal use and imported only 50,000 tons, by 1971 it produced less than 50,000 tons, and imported 400,000 tons — almost 90 percent of Colombia's domestic consumption.[5]

Free trade agreements in Stages 3 and 4 built on food aid dumping strategies of earlier stages with predictable results: local agricultural production was decimated and subsistence and small farm sectors that produced for local markets were particularly hard hit. This fueled social turmoil that was met with state-sponsored terror. Colombian human rights leader Héctor Mondragón says that it has been the standing policy of a long series of Colombian governments and their U.S. multinational corporate allies, that "in order for there to be development in Colombia it is necessary to get rid of the campesinos." "I think if the U.S. wants to do something good for Colombia," Mondragón says, "then the first thing they should do is change the trade policies with Colombia. Because primarily the free trade agreements and the World Trade Organization . . . have created a situation that has left the Colombian agricultural sector in ruins." "The real 'crop substitution' that has happened in Colombia," he notes, "is substituting poppy for coffee. And this is the fault of the WTO." He writes:

> The campesino doesn't have any other means by which to live. The other licit crops are not profitable anymore because of free trade. The importations have put anyone trying to produce licit crops into ruins. When the crops of a campesino are destroyed or ruined they don't have any alternative but to sell their land to a large landowner. Curiously, this land is generally sold to a drug dealer who has money. Then the small farmer goes deeper into the jungle to plant illicit crops. Before this farmer lived with 2 or 3 acres of coca, now they'll have to plant 4 or 5 be-

cause the cost of production if you are further away is higher. And so to be able to survive the small farmer cuts down more jungle and plants more crops. And for this reason the fumigation, rather than lowering the amount of acres cultivated, actually increases them. This is stupid policy. And so therefore one has to realize that the objective must not be to get rid of the illegal crops. The objective is the same as the war and the violence: displace the campesinos. Destroy their communities and organizations.[6]

Joseph C. Leuer has emerged as a kind of official defender and historian of the School of the Americas. One of his contradiction-laden defenses of the school is an information paper titled, "School of the Americas and U.S. Foreign Policy Attainment in Latin America," in which Leuer describes "the School's role as *a forum to establish stability through military-to-military engagement*."[7] This admission makes it abundantly clear that Latin American officers trained at the school are instruments of U.S. foreign policy. They were given the means and training to deal with problems such as displaced and disgruntled campesinos and to impose "stability" prized by U.S. political and business leaders through "military-to military engagement."

Predictably, as the U.S. tried to achieve stability in the absence of democracy, freedom, development, or human rights, its foreign policy in Stage 1 depended on brutal force and intimidation. Predisposed to use the generals to defend economic and strategic interests, U.S. foreign policy planners equipped School of the Americas' graduates with the repressive skills they needed to maintain stability in the midst of hunger, poverty, and social inequality that deepened during the generals' rule. "The U.S. public," Leuer writes, "has difficulty understanding past U.S. support of sometimes authoritarian regimes in view of reported human rights abuses that have been attributed to them by international human rights organizations. However, past governmental support for many oligarchic Latin American regimes was accepted as necessary...."[8] At one point Leuer claims such support was necessary "to stimulate the transition to participatory democratic governmental structures in a competitive bipolar world."[9]

Elsewhere he says more honestly that, following World War II, "foreign aid programs, specifically the Mutual Securities Act of 1951, began to tie foreign development aid directly to military aid and anti-Marxist allegiances. At the time," Leuer writes, *"the military was perceived as the only stable force which could achieve the U.S. goal of denying access to government by the revolutionary thinkers."*[10] Leuer then names and passes over the brutal contradictions at the heart of U.S. policy:

> U.S. analysts believed there were three major factors which would allow for communist expansion in Latin America: (1) Latin American resentment of U.S. intervention in the Americas, (2) Neoclassical development schemes imposed on Latin American governments by large U.S.-controlled multinational companies, and (3) U.S. support of Latin American elites tied directly to the repressive military structure and the United States.[11]

Little wonder that the School of the Americas, established in 1946 to promote "stability through military-to-military engagement," became known by other names: "School of Coups," "School of Assassins," and "School of Dictators." An editorial in the *Atlanta Constitution* urging closing of the SOA notes:

> Over the years, the school developed a reputation that was the opposite of stability. So many of its attendees played leading roles in overthrowing governments — Panama, Ecuador, Bolivia, Peru, and Argentina — that their alma mater came to be known as "la escuela de golpas" (the school for coups). A decade ago it acquired a more ominous nickname, the School for Assassins, from a Panamanian newspaper at the time Panama severed its links with the school. The label was to call attention to the unhappy penchant of the school's alumni to turn up as suspected death-squad officers throughout Central America.[12]

The U.S. created national security states throughout Latin America during Stage 1. A year after the opening of the School of the Americas in Panama, the U.S. Congress passed the National Security

Act of 1947. This act created the Central Intelligence Agency and
the National Security Council. The ethical grounding of these agencies, linked to frequent acts of deception, destabilization, and terror,
was the belief that the United States must use any means necessary
to defend vital interests. A secret government commission headed by
former president Herbert Hoover issued a report in 1954 calling for
the adoption of repressive tactics in the conduct of foreign policy:

> It is now clear that we are facing an implacable enemy whose
> avowed objective is world domination. . . . There are no rules in
> such a game. Hitherto accepted norms of human conduct do
> not apply. . . . If the United States is to survive, long-standing
> American concepts of fair play must be reconsidered. . . . We
> must learn to subvert, sabotage, and destroy our enemies by
> more clever, sophisticated, more effective methods than those
> used against us.[13]

That same year President Eisenhower presented the Legion of
Merit to two Latin American dictators — Pérez Jiménez of Venezuela for his "spirit of friendship and cooperation" and his "sound
foreign investment policies," and Manuel Odría of Peru. Nineteen
fifty-four was also the year the CIA carried out its notorious coup in
Guatemala to protect United Fruit and other U.S. companies from
labor unions and land reform, setting up forty years of almost unin-
terrupted terror. Thomas McCann, a former vice-president of United
Fruit, describes why his company found Guatemala attractive and
what went wrong:

> Guatemala was chosen as the site for the company's earliest de-
> velopment activities because a good portion of the country was
> prime banana land and also because at the time we entered
> Central America, Guatemala's government was the region's
> weakest, most corrupt, and most pliable. In short, the coun-
> try offered an "ideal investment climate," and United Fruit's
> profits flourished for fifty years. Then something went wrong:
> a man named Jacobo Arbenz became president.[14]

What went wrong in Guatemala from the perspective of U.S. political and business leaders was democracy. A democratically elected leader legalized labor unions, set up social security, and carried out a land reform program. This was unacceptable because, using Leuer's terms, a "revolutionary thinker" had been elected president of Guatemala. He did not remain president for long. United Fruit Company and other U.S. businesses had friends in high places. At the time of the CIA coup, John Foster Dulles, a long-time legal advisor to the company, was U.S. Secretary of State; his brother Allen Dulles was director of the CIA; Henry Cabot Lodge, a large stockholder and member of United Fruit Company's board of directors, was the U.S. ambassador to the United Nations; John Moors Cabot, a large shareholder, was Assistant Secretary of State for Inter-American Affairs; and Walter Bedell Smith, predecessor of Allen Dulles as CIA director, became president of United Fruit Company after the Arbenz government was overthrown.

Within the parameters of the Hoover Commission report and an "any means necessary" foreign policy, the CIA and its extensive network of assets, contacts, and agents became a sort of presidential hit squad sent out to defend corporate interests in the name of national security. The mission was "to subvert, sabotage, and destroy our enemies." The means used not only violated "hitherto accepted norms of human conduct," as the evidence against the SOA makes clear, but they often circumvented the law, the will of Congress, and the conscience and moral sensibility of the people. Despicable goals and means were shrouded in secrecy and the mythology of the benevolent superpower.

In *The Iran Contra Connection*, Peter Dale Scott, Jonathan Marshal, and Jane Hunter describe the origins of U.S. covert capabilities:

> From their inception to the present, many CIA operations have been covert, not just to deceive foreign populations, but at least partly because they were *designed* to violate U.S. statutes and Congressional will. A relevant example is the so-called "Defection Program" authorized in 1947 (by National Security Council Directive 4, a document still withheld in full). Despite

explicit congressional prohibitions, this program was designed to bring Nazi agents, some of them wanted war criminals, to this country to develop the covert capability of the United States.[15]

The Church and the National Security State

Once the United States established its own intelligence and covert operations capability, it set out to assist friendly military regimes throughout the Americas in setting up similar structures. The result, as José Comblin documents in his book *The Church and the National Security State*, was that throughout Latin America state power was exercised through military leaders and institutions. Comblin called such military-dominated nations national security states. Such states, he said, set legal or functional limits on constitutional authority; justified human rights and other abuses by agents of the state by appealing to higher values or defense of the state itself; sought national unity based on attacks against external or internal enemies, usually defined as communists; and positioned state power in military hands.[16]

National security states also expected and demanded that the church mobilize its financial, ideological, and theological resources in service to the state. Traditional theology had arrived with the conquest and generally had justified oppression and positioned the church as an institution of privilege and power serving the interests of the powerful. It explained poverty as God's will, riches as God's blessing, obedience to religious and political authorities as a requirement, and it offered the pacifying promise of a heavenly reward. Salvation was individual and the social causes of injustice were kept outside the parameters of theology, church concern, or religious action.

Changes in the Catholic church shook the foundations of the established order. Building on Vatican II (1962–65), which had called the church to be a living sign of God's loving presence in the world, the Latin American bishops (Medellín, Colombia, 1968) affirmed a theology of liberation in response to their world of poverty and misery. Liberation theology clashed sharply with traditional theology. It stressed that poverty reflected unjust systems and not God's will. It affirmed the dignity of the human person and said that in order for political, economic, and social structures to reflect God's intent they had to be organized to make dignified living possible. Liberation theology named concentrated land ownership, the huge gap separating the rich and poor, and other structural problems as examples of social sin and institutionalized violence. It spoke of a spiral of violence in which oppression of the poor sparked rebellion that in turn was met with repressive violence. The spiral could be broken only if the church worked to end oppression and broke its cozy relationship with the powerful military, economic, and political forces that controlled the oppressive social order. Liberation theology transformed Jesus from simply an object of faith to an example of faithfulness, from a passive victim to that of an executed subversive who, like faithful Christians today, challenged oppressive groups and systems. Jesus risked his life in service to justice and liberation and he was killed. The resurrection was God's vindication of his life and faith, which Christians and the church were called to exemplify.

The military despised liberation theology and those living out the new theology were met with a wave of violence. Penny Lernoux describes how organized attacks against progressive religious escalated throughout Latin America as new ways of being church took root. At the center of the strategy of religious persecution was SOA graduate and Hall-of-Fame member Hugo Banzer:

> Archbishop Alfonso López Trujillo ... denounced a "concerted campaign against the church." While there were plenty of skirmishes between national churches and governments in Latin America during the early 1970s, only since 1975 has there been a marked similarity in these anti-Church campaigns, such as planting communist literature on Church premises or arresting foreign priests and bishops on trumped-up charges of "subversion." This was no coincidence: at least eleven countries were following the same geopolitical plan, the Doctrine of National Security, and therefore shared similar strategies in their "war on communism."

A typical contribution to this common strategy was the "Banzer Plan," hatched in the Bolivian Interior Ministry in early 1975 and named for Hugo Banzer, Bolivia's right-wing military dictator. Although the original plan was not committed to paper, it was discussed at length in the Interior Ministry, a publicly acknowledged subsidiary of the CIA, after the Bolivian Church began to make trouble for the government by denouncing the massacre of tin miners. The plan was leaked to Bolivia's Jesuits by an Interior Ministry official, who was horrified by the government's intention to smear, arrest, expel, or murder any dissident priest or bishop in the Bolivian Church. The authenticity of the Banzer Plan, which boasted many of the classic "dirty tricks" employed by the CIA in Latin America during the 1960s, was subsequently confirmed by the government itself when it followed, word for word, all the original tactics. . . . The three main thrusts of the campaign were to sharpen internal divisions within the Church, to smear and harass progressive Bolivian Church leaders, and to arrest or expel foreign priests and nuns, who make up 85% of the Bolivian clergy.[17]

Jon Sobrino, reflecting on the murder of his fellow Jesuits in El Salvador, notes the high price paid by those religious who overstepped the boundaries established by the national security state:

Women and men have shed their blood — people from El Salvador, from Spain, and from the United States. People from different confessions, from different faiths, from different places, are united in their soul, as we all are by the tragedy in El Salvador and also by the hope and the commitment of the martyrs. We all know why these people ended in the cross, why they were killed. They dared "touch the idols of death." . . . Archbishop Romero of El Salvador defined the idols as the accumulation of wealth and the doctrine of national security. Those who dare touch these idols get killed.[18]

Full-fledged national security states, and the ideology that guided their function and formation, did not emerge fully in Latin America until after the anti-communism hysteria that followed the Cuban revolution in 1959. Their seeds were planted and watered by U.S. policies much earlier. The content and recommendations of the Hoover Commission; the bestowal of Legion of Merit awards to dictators; the CIA-led overthrow of a democratically elected government in Guatemala; Kennan's dictate that the United States could not afford to be altruistic and therefore should cease to talk about human rights, democracy or development; "Defection Directives" bringing Nazi war criminals to the United States to help us establish intelligence operations; exporting similar organizations and the national security state ideology to justify their repressive tactics; alliances with military despots; and training soldiers in repressive tactics at the School of the Americas are all part of a hidden history unknown to most U.S. citizens who are deeply socialized into the comforting myth of the benevolent superpower.

In the aftermath of the Cuban revolution all of these repressive impulses and the mechanisms for actualizing them deepened. The U.S. responded to the overthrow of the U.S.-backed dictator Fulgencio Batista with the Alliance for Progress, a program with conflicting and irreconcilable impulses and components. On the one hand, promoters of the Alliance recognized that poverty was a breeding ground for social turmoil. This led to a slight increase in economic aid and a large increase in reform rhetoric. On the other hand, the Alliance increased support dramatically for Latin American militaries and the focus of aid and training shifted further in the direction of counterinsurgency warfare against internal enemies.

Any expectations of economic reform and improved living conditions generated by the Alliance for Progress were dashed quickly. Preoccupations with stability, defense of corporate interests, and fear of internal enemies translated into increased U.S. support for repressive militaries. Senator Edward Kennedy summarized the disastrous results:

> Their [Latin American countries'] economic growth per capita is less than before the Alliance for Progress began; in the previous eight years U.S. business has repatriated $8.3 bil-

lion in private profits, more than three times the total of new investments; the land remains in the hands of a few; one-third of the rural labor force is unemployed and *13 constitutional governments have been overthrown since the Alliance was launched.*[19]

During Stage 1 of U.S. foreign policy repressive militaries and their security organizations stood at the pinnacle of power. They defended elite economic interests, equated democracy with subversion, and justified repression in the name of fighting dangerous internal and external enemies. Students, peasants, unionists, and progressive religious were considered local representatives of an international communist conspiracy. Seeking a stable investment climate divorced from human rights, democracy, and development inevitably led to a School of the Americas steeped in repressive tactics in service to a foreign policy marked by violence, persecution, and terror. The terror would escalate in many places during the 1980s but along with the repression one could see bankers in three-piece suits carrying out U.S. foreign policy with similarly dire consequences.

Chapter 7

Geopolitics and the SOA/WHISC

Foreign Policy Stages 2–4

The 1980s were the decade of Reaganomics, a period within the United States marked by large tax cuts and wealth gains for the richest Americans, unprecedented inequality, huge increases in military spending, and ballooning national debt. Soaring budget deficits, the inevitable and conscious result of policy, were used to justify social cutbacks, force restructuring of the economy, and reduce the role of government.

Stage 2 of U.S. foreign policy (roughly 1980–90) extended many of the values and priorities of Reaganomics into international affairs. The powerful economic groups that shaped domestic policies to their own benefit extended their vision and interests into foreign affairs as domestic and foreign policy goals converged in the 1980s and beyond. As a result of U.S. policies, wealth was concentrated further, military power and budgets increased, and debt was used to leverage desired economic and political outcomes both within the United States and throughout Latin America. Poor country elites prospered while poor majorities suffered as the U.S. provided "third world" militaries and paramilitaries, wherever deemed necessary, with enormous financial resources, training, and ideological cover for waves of repression. At the same time, or in other settings, debt was used to force third world nations to implement "structural adjustment programs" that imposed conditions favorable to foreign investors, including social cutbacks, privatization, increased exports, and labor discipline.

Two Tracks

Stage 2 of U.S. foreign policy was carried out on two tracks. Track 1, evident throughout Central America in the 1980s, deepened the repressive violence featured centrally in Stage 1. The U.S. created, trained, and financed the contras who, under U.S. tutelage, engaged in tactics of terror and torture in an effort to destabilize the revolutionary government of Nicaragua. Nicaragua's main "crime" was not that it was a Soviet beachhead as charged by U.S. leaders, but that it was trying to find a viable alternative for a third world country that would break from the molds of either repressive U.S.-style capitalism or repressive Soviet-style socialism.

In El Salvador the U.S. backed a series of repressive governments, supported death squads, and orchestrated a bloody war against popular sectors, including progressive elements of the church. A political officer at the U.S. Embassy in El Salvador explained to me in a fit of rage why the United States encouraged terror tactics in the conduct of the war. "All right," he said, his voice rising and his cheeks flushed red, "what is it that troubles you most? You didn't like the massive repression of several years ago when dozens of bodies appeared every day in the street. Well, if you didn't like massive terror," he continued, "remember this: We had no other choice." Each word was spoken slowly, clearly, and loudly like a parent reprimanding a naughty child. "Nicaragua," he explained, "had already fallen and a massively popular revolution was about to take power in El Salvador. We had no other choice." Then he stopped and regained his composure. A disturbing moment of truth-telling had passed. His expressed rage had swirled around the embassy room like an uninvited demon shedding light on a dirty secret.

Track 1 of Stage 2 reflected the grisly obscenity of an "any means necessary" foreign policy. I have written elsewhere about U.S. low intensity warfare strategy and its component of managing terror to fit particular political circumstances.[1] Here I want to note that many of the most gruesome abuses linked to the School of the Americas and its graduates were committed during Stage 2, including the murders of Romero, the four U.S. churchwomen, and the Jesuit priests in El

Salvador. When some of these abuses came to light as a result of investigations by Congressman Moakley, the United Nations Truth Commission, and the research and witness of SOA Watch, a wave of editorials calling for the school's closing followed.

The *San Antonio Express* called the SOA a "breeding ground for human rights abusers."[2] The *Des Moines Register* said the SOA serves to "advance the education of killers."[3] The *Atlanta Constitution* noted that the decision to close the SOA "should be an easy one" because it "has strung together such a perverse 'honor roll' of cold-blooded murderers that America's meanest prison might be pressed to match it."[4] The *Cleveland Plain Dealer* editorialized that the "SOA's best known products have shared a distressing tendency to show up as dictators or as leaders or members of death squads. They have been agents of oppression."[5] The *New York Times* editorialized that closing the School of the Americas would "make it clear that in the future rogue operators, who abuse their relationship with the United States, will be exposed rather than protected." It would "announce that America will no longer train and encourage Latin American thugs."[6]

What these editorials fail to say completely and honestly is that these "thugs," "human rights abusers," "cold-blooded murderers," "dictators," and "leaders or members of death squads" were not "rogue operators" who abused "their relationship with the United States." They were *valued operatives carrying out U.S. foreign policy.*

Track 2: Any Means by Other Means

If Track 1 of Stage 2 demonstrated the degree of terror and repression that U.S. foreign policy was capable of promoting, then Track 2 of Stage 2 showed flexibility and creativity. Stage 2 of U.S. foreign policy functioned within the framework of low-intensity conflict (LIC). LIC integrates military, psychological, diplomatic, and economic aspects of warfare into a comprehensive package. Track 2 of Stage 2, however, elevated economic aspects of warfare to an unprecedented degree, especially the use of debt as leverage. This meant that in the 1980s the International Monetary Fund (IMF) and

World Bank (WB) became important, and in many cases, the principal instruments of U.S. foreign policy as they imposed structural adjustment programs (SAPs) on vulnerable peoples. The Ecumenical Coalition for Economic Justice describes the impact of SAPs on third world countries:

> Instead of developing their own resources to meet pressing human needs, many Third World economies are literally being "sapped" — gradually exhausted of their wealth — through conditions imposed by their creditors. The goals of this new colonialism are, in part, the same as the old. Thanks to SAPs, transnational corporations enjoy greater access to cheap raw materials, cheap labour and foreign markets. But... the contemporary recolonization also involves an annual collection of tribute in the form of interest payments on debts that... can never be paid off. Thanks to the "success" of SAPs, debt bondage is becoming permanent.[7]

The standard features of IMF structural adjustment programs include: currency devaluations; higher interest rates; strict control of the money supply; cuts in government spending; removal of trade and exchange controls; the use of market forces to set the prices of goods, services, and labor; privatization of public sector enterprises; and export promotion. These measures, according to IMF theory, should result in lower inflation, increased exports, reduced consumption and imports, greater efficiency, international competitiveness, and substantial foreign exchange earnings available for debt servicing.

SAPs in practice have been a dramatic success for first world elites and a deadly failure for the third world poor. In Stage 2 the IMF was assigned a vital and conscious role as an instrument of U.S. foreign policy. As a result, indebted countries were forced to restructure their economies to meet U.S. determined standards and goals. The IMF-led foreign policy achieved three objectives for U.S. businesses.

First, IMF policies ensured a continuous transfer of wealth to rich countries through interest payment on third world debt. Between 1982 and 1989, for example, the net outflow of debt service from underdeveloped to developed countries, that is, the amount of capital exported in excess of new loans, equaled $240 billion dollars.[8] Despite this massive transfer of wealth from the poor to the rich, the World Bank reported that in the five years after 1982 no country rescheduling debts actually reduced the ratio of that debt to its gross national product.[9]

A second objective achieved via IMF SAPs is that conditions imposed force third world nations to integrate their economies into the international system on terms favorable to the United States. Finally, and a corollary to the second point, IMF policies encourage greater foreign penetration and control of third world resources and economies. SAPs hurt poor countries in at least seven ways:

- The required emphasis on export production often weakens the subsistence sector while strengthening sectors dominated by foreigners.

- When the export requirement is imposed on numerous countries at the same time it can result in overproduction and a further deterioration in the terms of trade.

- Higher interest rates mandated by the IMF often encourage speculation, fuel inflation, and aggravate class divisions as credit is limited to the most affluent and powerful economic actors.

- Removal of trade and export controls fosters dependence on foreign imports, increases the domination of foreign firms over domestic ones, and encourages capital flight.

- Privatization required by the IMF can result in greater concentration of wealth and a loss of economic sovereignty.

- Mandated currency devaluations erode the purchasing power of workers while benefiting foreign corporations operating in export zones.

- Finally, satisfying the IMF and foreign creditors requires third-world governments to drastically reduce government spending.

The lessons of Track 2 of Stage 2 are that in many settings the IMF can be as effective an instrument of U.S. foreign policy as contingents of SOA-trained soldiers and, for poor majorities, these policies can be as deadly as any bullet and as painful as any torturer's hand. According to the United Nations children's organization (UNICEF) the world's thirty-seven poorest countries cut healthcare budgets by fifty percent and education budgets by twenty-five percent in the 1980s. UNICEF estimates that more than a million African children died in the 1980s as a result of structural adjustment programs imposed on the poor. In 1988 alone, according to UNICEF, 500,000 children died in underdeveloped countries as a direct result of SAP-induced austerity measures. UNICEF's negative assessment of SAPs is blunt and startling:

> It is essential to strip away the niceties of economic parlance and say that . . . the developing world's debt, both in the manner in which it was incurred and in the manner in which it is being "adjusted to," . . . is simply an outrage against a large section of humanity.[10]

SAPs hurt the poor and they fail to achieve officially stated goals. From the perspective of U.S. foreign policy objectives, however, they are a stunning success. SAPs set limits on economic policy options and restrict political debate, ideas, and action. These anti-democratic tendencies, as discussed below, have deepened as corporate-led globalization has become more institutionalized during Stages 3 and 4 of U.S. foreign policy. Today, to insure stability and establish a favorable investment climate, assignments given previously to dictators and repressive militaries are *whenever possible* left in the hands of bankers.

During Stage 2 of U.S. foreign policy the IMF and World Bank functioned as an economic police force that worked with great efficiency worldwide on behalf of elite economic clients. The Ecumenical Coalition for Economic Justice offers the following summary of costs and benefits of SAPs:

> Given the evidence that SAPs do not achieve their official goals, that they cause immense hunger and misery and they accentuate underdevelopment, why do private bankers, the

IMF, the World Bank, and conservative governments insist on their strict application?...Viewed from the perspective of transnational investors, SAPs do make sense. SAPs assure transnational corporations that countries on the periphery will supply abundant supplies of cheap raw materials, low-wage labour, and markets for some of their products. SAPs enable transnationals to maintain control over manufacturing processes, technology, and finance, sharing some of the spoils with local elites. In addition, SAPs promote exports that earn foreign exchange to service otherwise unpayable debts.[11]

In Stage 1 of U.S. foreign policy and within Track 1 of Stage 2 repressive militaries were called on to insure stability and favorable investment climates for elite corporate interests. In Track 1 of Stage 2 repressive violence rose to new heights as terror and torture tactics reflected the "any means necessary" philosophy that guided and *continues to guide* U.S. foreign policy. The School of the Americas played a vital role in carrying out the bloody repression.

Track 2 of Stage 2 relied on different tactics and instruments. Track 2 was based on the realization that dictators and military despots could be destabilizing rather than guardians of stability and the understanding that "any means necessary" didn't necessarily mean increased overt violence. The ultimate goal of foreign policy is to get others to do what you want them to do. U.S. foreign policy, using a variety of means and tactics, has always been the instrument by which powerful economic actors use levers of state power to insure that other nations set economic rules and conditions agreeable to U.S. interests.

Track 1 of Stage 2 was steeped in repressive violence. In settings such as El Salvador and Nicaragua brutal violence was seen as the necessary and perhaps only effective means of power projection. Track 2 was based on the knowledge that indebtedness made third world nations vulnerable and gave the U.S. enormous power. The United States can at times, in other words, achieve economic and strategic objectives as or more effectively by sending in bankers instead of repressive military forces. Militarization and repression are sometimes necessary, according to this view, but are best viewed as secondary options. In Track 2 the United States imposed IMF and World Bank structural adjustment programs in order to insure that Western banks were paid, to transition from dictatorship to restricted democracy, and to establish firm rules for the international economic game. Third world nations were and are invited and even forced to play, but they cannot make the rules. They are often given the option to play or be destroyed.

Most third world governments, themselves representing elite interests, play the game and place the burdens of adjustment on the poor. They join forces with the IMF to reduce social spending, privatize economies, allow profits and capital to flow freely, keep labor costs down and unions weak, and increase exports at the expense of domestic consumption and the environment. They cooperate because they have few realistic choices and because their power is tied to foreign economic interests. "The third world elites who borrowed the money," Jorge Sol, a former IMF executive director for Central America, states, "come from the same class as those who lent it and as those who managed it at the IMF. They went to the same schools, belonged to the same clubs. They all profited greatly from the debt."[12] "Once a country makes the leap into the system of globalization," Thomas Friedman writes approvingly, *its elites begin to internalize this perspective of integration, and always try to locate themselves in a global context.*"[13]

Stages 3 and 4: The Primacy of Economic Power

Track 2 of Stage 2 elevated economic leverage and economic power as instruments of foreign policy. In many cases this led to a reduced role for the U.S. and Latin American militaries in the 1990s. This is an important factor in assessing the present role of the SOA/ WHISC (see Chapter 9).

Stage 3 of U.S. foreign policy (roughly 1991–97) is marked by two important dynamics. First, both Stages 3 and 4 reflect new geopolitical realities. By the early 1990s the Cold War was over. The Soviet Union disintegrated and the "second world" disappeared. Nicaragua had been effectively destroyed as part of U.S. "low-intensity conflict" strategy and El Salvador was simply exhausted. Nicaragua and El Salvador, tiny countries that U.S. foreign policy planners declared the most important places on earth in the 1980s, returned to their previous status as insignificant drops in a mighty sea. Most of their war-weary people were abandoned and left to drown. It was now a unipolar world. The United States was the lone superpower and it was poised to expand its influence and power with globalization rhetoric replacing outdated anti-communism. As Thomas Friedman notes, in "the globalization system, the United States is now the sole and dominant superpower and all other nations are subordinate to it to one degree or another."[14]

In the unipolar world of corporate-led globalization all nations, particularly third world ones, have nowhere to turn and options for all but the most powerful are limited to playing by the established rules of the global economy or not playing at all. Friedman writes:

> ...on the political front, the Golden Straitjacket narrows the political and economic policy choices of those in power to relatively tight parameters. That is why it is increasingly difficult these days to find any real differences between ruling and opposition parties in those countries that have put on the Golden Straitjacket. Once your country puts on the Golden Straitjacket, its political choices get reduced to Pepsi or Coke — to slight nuances of taste, slight nuances of policy, slight alterations in design to account for local traditions, some loosening here or there, but never any major deviation from the core golden rules.[15]

The Golden Straitjacket must be worn by rich and poor nations. "No wonder," Friedman writes, "so much of the political debate in developed countries today has been reduced to arguments over minor tailoring changes in the Golden Straitjacket, not radical alteration."[16] This undermining of democracy, according to Friedman, is a necessary and worthwhile trade-off because only the Golden Straitjacket, the Electronic Herd, and the Supermarkets of corporate-led globalization can deliver prosperity.

Track 2 of Stage 2 of U.S. foreign policy used debt as leverage and imposed structural adjustments on poor nations. This laid the foundations for Stages 3 and 4. Track 2, sometimes in the aftermath of the bloody violence of Track 1, established the limited parameters within which nations and governments could make decisions about political and economic life. Theologian John Cobb describes the policies and priorities demanded by what he calls the "Washington Consensus":

> The Washington Consensus centers on the idea that development should be through corporate investment rather than governmental and intergovernmental grants and loans. To encourage corporate investment, national boundaries should be downgraded so that goods and capital can flow freely around the world. The ideal is that a single global market replace the international economy of earlier decades. Each country should privatize all productive assets and make them available for purchase by international capital. It should cease to protect its businesses from international competition and concentrate only on that production in which it has a comparative advantage. It should adopt governmental policies that make it attractive for transnational investment. To a truly remarkable degree, the Washington Consensus has governed global economic policy since the early 1980s, and in the nineties it has approached its goal of a unified global market.[17]

The policies and practices of the "Washington Consensus," the "Golden Straitjacket," the "Electronic Herd" and the "Supermarkets" restrict freedoms. You are a government that wants to spend more money on education. You can't. You want to subsidize the cost of basic foods. Sorry, IMF structural adjustment conditions prevent you from doing so. You want to implement agricultural policies that help subsistence and other local farmers to insure a measure of food security and to discourage migration to urban slums. You can't because such policies violate the rules of free trade established through NAFTA and the WTO. You want to protect your environment. You can't because you need to attract foreign investors and expand exports and because your nation's resources are no longer your nation's resources. They are part of the massive pool of raw materials that are available to anyone of any nation with capital. Any laws you pass in an effort to restrict unimpeded access will be judged as violations of free trade and overturned by WTO enforcers.

Repressive militaries were the main obstacle to democracy during Stage 1 and within Track 1 of Stage 2 of U.S. foreign policy. Since the end of Stage 2 and within Stages 3 and 4 the *principal* obstacle to democracy both at home and abroad is concentrated economic power and the corresponding consolidation of decision-making power in the hands of economic elites. The "main job" of all world leaders, Thomas Friedman writes, "is enticing the Electronic Herd and Supermarkets to invest in their states, doing whatever it takes to keep them there and constantly living in dread that they will leave."[18] According to Friedman:

> ... the Electronic Herd and the Supermarkets are fast becoming two of the most intimidating, coercive, intrusive forces in the world today. They leave many people feeling that whatever democracy they have at home, whatever the choices they think they are exercising in their local or national elections, whoever they think they elected to run their societies, are all just illusions — because it is actually larger, distant, faceless markets and herds that are dictating their political lives. The paradox of globalization is that some days the herd rides into town like the Lone Ranger, guns blazing, demanding the rule of law, and the next day it stomps right out of town like King Kong, squashing everyone in its path.[19]

The IMF/World Bank policies imposed in the 1980s limited economic decisions throughout the third world to a narrow range of acceptable options and determined the "proper role" for government. Domestically, Reaganomics did much the same. As Friedman acknowledges, Britain's Margaret Thatcher and Ronald Reagan "combined to strip huge chunks of economic decision-making power from the state, from the advocates of the Great Society and from traditional Keynesian economics, and hand them over to the free market."[20]

Although President Clinton wanted a more activist government to address some of the social problems that result from corporate-led globalization, he was a strong advocate of the corporate agenda. During Stage 3, then, President Bill Clinton defined the North American Free Trade Agreement (NAFTA) as his principal *foreign policy* issue. Stages 3 and 4 of U.S. foreign policy are marked by accelerated efforts to institutionalize a corporate-led global economic order whose foundations were firmly established through the IMF structural adjustment programs and military repression of earlier stages. Through institutions such as the IMF, through trade agreements such as NAFTA and the FTAA, and through supra-corporate rule-making bodies such as the WTO, corporations are today the principal instruments and beneficiaries of U.S. foreign policy. Global institutions such as "the UN, the World Bank, and the IMF," Friedman writes, "are critical for stabilizing an international system from which America benefits more than any other country."[21] "There are big, important places and there are small unimportant places, and diplomacy is about knowing the difference between the two, and knowing how to mobilize others to act where we cannot or should not go," Friedman writes. "The very reason we need to support the United Nations and the IMF, the World Bank, and the various world development banks is that they make it possible for the Untied States to advance its interests without putting American lives on the line everywhere, all the time."[22] One could make nearly the same argument in defense of the SOA/WHISC.

Friedman's position with the *New York Times* itself reflects the shift in U.S. foreign policy in Stages 3 and 4 away from military power projection to economic power projection. His new position was created in 1994 to *"cover the intersection between foreign policy and international finance."*[23] He describes how economic policy planners now dominate foreign policy:

> [S]trategizing has been done by people whom you don't usually associate with grand strategy. Their names are Greenspan [head of the Federal Reserve], Rubin [head of the Treasury Department] and Summers [head of the World Bank]. But

don't think just because they did the strategizing, and not the Secretary of State or the Secretary of Defense, that it does not require a global vision and has not put in place global structures that will fundamentally shape, and hopefully stabilize, relations between states. If that isn't grand strategy, then I don't know what is. If that isn't foreign policy, then I don't know what is.[24]

Economic organizational structures created and free trade agreements negotiated during Stages 3 and 4 of U.S. foreign policy allow corporations to institutionalize their power and impose their agenda worldwide, a task previously assigned most directly to the U.S. military and its military allies.

Stage 3: Military Downsizing

A second important dynamic within Stage 3 of U.S. foreign policy is that Latin American militaries and the U.S. military-industrial-congressional complex were put in an awkward position by the collapse of the Soviet Union and the success of economic leverage as a determining factor in U.S. foreign policy. Col. Patricio Haro Ayerve, subcommandant of the School of the Americas, wrote shortly before the school's official name change:

> The environment in which the School of the Americas was established has gradually changed over the years, as the post-Cold-War era and unipolarity have produced a much different scenario which, consequently, explains why the countries of the world have been forced to redefine and redirect their security issues. Therefore, the emergence of new and different threats to nations and the security system of their inhabitants make it necessary to train the armed forces and police officers of the Americas to deal with these new concepts.[25]

Track 1 of Stage 2 of U.S. foreign policy escalated military violence wherever necessary. Track 2 of Stage 2, however, included the frequent and effective exercise of power through economic leverage. Stages 3 and 4 were built on the foundations of these successes. They elevate the role of economic power as determinative in U.S. foreign policy. They institutionalize gains made possible by past repression and successful IMF/World Bank structural adjustments. The rules and conditions required by the IMF have achieved binding legal status through organizational structures, rules and treaties linked to NAFTA and the WTO.

The emergence of economic leverage as a foreign policy tool in Stage 2 and the ascendancy and institutionalization of economic power in Stages 3 and 4 left the U.S. military and its third world military allies in trouble and without a mission. The problem was twofold. The Soviet Union, the principal justifier of inflated "defense" budgets, was no more; and, economic power, institutionalized and effective in the unipolar world of corporate-driven globalization, made the U.S. military and the repressive militaries it supported less necessary, even counterproductive. This dynamic, as we will see in Chapter 9, has profoundly impacted the mission and purpose of the SOA/WHISC. Globalization, Friedman writes, "creates new sources of power, beyond the classic military measures of tanks, planes and missiles, and it creates new sources of pressure on countries to change how they organize themselves, pressures that come not from classic military incursions of one state into another, but rather by more invisible invasions of Supermarkets. . . ."[26]

We shouldn't feel too sorry for downsized militaries. The U.S. military-industrial-congressional complex is powerful and resilient and is presently upsizing. And Latin American counterparts are finding their place in the geopolitical environment of globalization. Larry Rohter in a *New York Times* article, "Latin America's Armies Are Down but Not Out," writes: "Civilian defense ministers in a lot of these countries don't order the generals around, they consult and negotiate with them." Latin American militaries, are, in other words, still a formidable power. In Central America, Rohter notes, the militaries have been downsized in terms of bullets and body counts but were well positioned to assert themselves economically:

> The Central American military offers a different but no less disturbing model. With no more guerrillas left to battle — or

slaughter — they have transformed into economic conglomerates that threaten to choke off competition from private enterprise. In a study titled "Soldiers as Businessmen," Arnoldo Brenes and Kevin Casas have documented how Central America's armed forces have taken over banks, hotels, funeral homes, radio stations, advertising agencies, supermarkets and stores through pension funds.[27]

Stage 4: Consolidation and Remilitarization

Stage 4 of U.S. foreign policy (1998 to the present) involves further consolidation of economic power and the remilitarization of foreign policy. There are three dynamics at play. First, new rounds of global negotiations aim to further concentrate economic decision-making power in the hands of large corporations. As we move through the various stages of U.S. foreign policy we see greater institutionalization and consolidation of economic rules and enforcement powers, from IMF-imposed structural adjustment, to NAFTA, to creation and strengthening of the WTO, to efforts to extend NAFTA throughout the Americas through the Free Trade Area of the Americas (FTAA).

Second, resistance to corporate-led globalization is exploding within the United States and throughout the world. Corporate led globalization is fracturing the world politically, culturally, economically, and environmentally (see Chapter 8). Resistance is building. This has prompted U.S. leaders to increase military spending and weapons sales and, most importantly, to expand dramatically U.S. military training of forces around the world. Friedman, whose symbol for "the drive for sustenance, improvement, prosperity and modernization — as it is played out in today's globalization system" is the Lexus and whose symbol of "everything that roots us, anchors us, identifies us, and locates us in this world" is the olive tree, writes:

> The biggest threat to your olive tree is likely to come from the Lexus — from all the anonymous, transnational, homogenizing, standardizing market forces and technologies that make

up today's globalizing economic system. There are some things about this system that can make the Lexus so overpowering it can overrun and overwhelm every olive tree in sight — and this can produce a real backlash.[28]

Toward the end of the book he warns:

Sustainable globalization requires a stable power structure, and no country is more essential for this than the United States. All the Internet and other technologies that Silicon Valley is designing to carry digital voices, videos and data around the world, all the trade and financial integration it is promoting through its innovations, and all the wealth this is generating, are happening in a world stabilized by a benign superpower, with its capital in Washington, D.C. The fact that no two countries have gone to war since they both got McDonald's is partly due to economic integration, but it is also due to the presence of American power and America's willingness to use that power against those who would threaten the system of globalization — from Iraq to North Korea. The hidden hand of the market will never work without a hidden fist. McDonald's cannot flourish without McDonnell Douglas, the designer of the U.S. Air Force F-15. And the hidden fist that keeps the world safe for Silicon Valley's technologies to flourish is called the U.S. Army, Air Force, Navy and Marine Corps. And these fighting forces and institutions are paid for by American taxpayer dollars.[29]

The United States accounts for approximately 45 percent of all weapons sales and it trains soldiers in at least 110 countries through the JCET (Joint Combined Exercise Training) program alone. There is also a renewed and direct relationship between U.S. militarization and oil, especially in Colombia, a nation where U.S. foreign policy today bears all the markings of the repressive tactics of El Salvador during Stage 1 and Track 1 of Stage 2. We should not be surprised that the largest number of soldiers trained over the past several years at the SOA/WHISC are from Colombia where we see overwhelming links between SOA graduates and human rights atrocities.

Finally, the third dynamic within Stage 4 is that remilitarization of U.S. foreign policy is driven by the needs and power of the military-industrial-congressional complex itself. Within Stages 3 and 4, more than at any other time in our nation's history, U.S. military spending and policy have been divorced from any credible threat or vital security need. Economic mechanisms for power projection in a unipolar world dominated by the United States dramatically reduce the need for high military budgets and would undercut efforts for aggressive militarization were it not for the undue influence of self-interested parties. The U.S. invasion of Panama, the Gulf War, the expansion of NATO, accelerated weapons exports, the escalation of the drug war, and missile defense systems must all be viewed in the context of a military-industrial-congressional complex desperate for enemies on which inflated budgets and wasteful military production depend.

In the aftermath of the Cold War, defense industry journals and military service reports reflected a palpable sense of anxiety, even panic. The Cold War thaw offered the possibility that hundreds of billions of dollars could be shifted from the military to other uses. A substantial peace dividend seemed likely. As a result, there was a desperate search among the various military branches and weapons makers for institutional legitimacy and for an ongoing role in a world that had unexpectedly changed.

Several examples can illustrate this point. A section of the May 1990 issue of the *Marine Corps Gazette* titled "On the Corps' Continuing Role" noted:

> The world is "in the midst of historic and promising trans-formation in the global security environment," experiencing changes more sweeping than any since the outbreak of World War II.... That the Cold War is over and the threat of global conflict has greatly diminished are widely held perceptions. Hopes for a "peace dividend" continue to be voiced, and major cutbacks in defense expenditures are regarded as virtual certainties.

Changes of this magnitude bring periods of "agonizing reappraisal." Clearly, the United States is entering a new era of reexamination of its defense needs. Policy, strategy, Service roles and functions, force structure, weapons systems, and budget levels all come under serious review. . . . The purpose of this section is to help put this challenging period in perspective and to encourage thinking about the Corps' future . . . [to] detail the specific capabilities that Marine Corps forces possess across the spectrum of warfare . . . [and to consider different] approaches the Corps might consider as it comes to grips with harsh budget realities.[30]

A few months later, General Colin Powell, following the collapse of the Berlin Wall and on the eve of the 1990 invasion of Panama, indicated that the U.S. military still had an important role to play. It needed to send a message of intimidation to the world: "We have to put a shingle outside our door saying 'superpower lives here' no matter what the Soviets do, even if they evacuate from all of Eastern Europe."[31] Foreign policy analyst James Petras saw a similar dynamic at play in the Gulf War in 1991. The war, he said, was meant to "intimidate the Third World into submission."[32] Within the Pentagon, Andrew and Leslie Cockburn note, the Gulf War was promoted as a calculated means of foreclosing on a peace dividend that was both possible in the context of the collapse of communism and dangerous to the interests of the military-industrial-congressional complex:

Short-term domestic political considerations aside, there were very important institutional imperatives behind the push toward military confrontation in the Gulf. . . . In April 1990 [months before the Iraqi invasion of Kuwait] a seasoned Pentagon official lamented in casual conversation that the atmosphere at his place of employment was dire. "No one knows what to do over here," he sighed. "The [Soviet] threat has melted down on us, and what else do we have? The Navy's been going up to the hill to talk about the threat of the Indian Navy in the Indian Ocean. Some people are talking about the threat of the Colombian drug cartels. But we can't keep a $300 billion budget afloat on that stuff. There's only one place that will do as a threat: Iraq." Iraq, he explained, was a long way away, which justified the budget for military airlift. It had a large air force, which would keep the United States Air Force happy, and the huge numbers of tanks in Saddam's army were more than enough to satisfy the requirements of U.S. ground forces.[33]

The authors note that Operation Desert Shield was so successful that it was known within the Pentagon as Budget Shield.[34]

I should also note that the drug war is a political, social and military disaster but it has given the U.S. military and its Latin American counterparts new life, new weapons, and new budgets. "Drug war politics has created a sprawling, generously funded 'narcoenforcement complex' of more than 50 federal agencies and bureaus," Coletta Youngers writes. "Despite the lack of progress in stemming illicit drug production and consumption, Congress continues to reward the narocenforcement complex generously."[35] Larry Rohter describes the implications of the drug war in Latin America:

One big question for civilian government [in Latin America] is what to do to keep the idled troops out of mischief. Washington's answer, regarded as a recipe for disaster by every government in the region, has been to press the armies to enlist in the war against drugs. "Americans tend to think that Latin America spends so much on the military that it might as well get something out of them," said Alfred Stepan, author of several books on the Latin American military. "But that is a dangerous way of looking at things. If the military get involved in the anti-drug wars, not only are they not prepared for it, but doing so has a corrupting impact and may contribute to an expansion of their role."[36]

The missile defense system is also driven by the needs of the military-industrial-congressional complex. It could turn out to be the biggest disaster for the planet and the biggest bonanza for the military-industrial-congressional complex in the history of the world. The new Bush administration is staffed by old Cold Warriors, including Vice President Dick Cheney, Secretary of Defense Donald Rumsfeld, and Otto Reich, an inside player in the Iran-contra scandal and now Assistant Secretary of State for Western Hemispheric Affairs. Political analyst Cedric Muhammad writes about Vice President Dick Cheney:

> There was no bigger supporter of the unbridled growth of U.S. military spending and the exporting of U.S. weapons of war than Dick Cheney. Far from a silent partner, Cheney advocated and participated in the proliferation of massive amounts of weapon sales in the Middle East. Cheney led the pack in the use of "arms sales" diplomacy in order to cobble together the anti-Iraq coalition in the Gulf War — offering nations in the region weapons in return for their support in getting Saddam Hussein out of the way.[37]

Defense Secretary Rumsfeld has extensive ties to right-wing "think tanks" and weapons contractors committed to implementing a missile defense system.[38] The Missile Defense system is unworkable and costly. "Since Reagan's 1983 speech [on Star Wars]," William Hartung writes, "we have spent over $70 billion on missile defense with virtually nothing to show for it."[39] Development and deployment of missile defense will set off a new global arms race and that may in fact be its intent. Hartung, of the World Policy Institute, wrote in *The Nation:*

> Under the guise of revising nuclear policy to make it more relevant to the post-Cold-War world, the Bush administration is pushing an ambitious scheme to deploy a massive missile defense system and develop a new generation of nuclear weapons. If fully implemented, Bush's aggressive new policy could provoke a multi-sided nuclear arms race that will make the US-Soviet competition of the cold war era look tame by comparison.... The President and his Star Warrior in Chief, Defense Secretary Donald Rumsfeld, are willing to put missile interceptors on land, at sea, on airplanes, and in outer space in pursuit of continued US military dominance.... But

even Washington's closest NATO allies continue to have grave reservations about Rumsfeld's suggestion that the United States might trash the Anti-Ballistic Missile treaty of 1972 in order to pursue its missile defense fantasy.... The cost of Bush's Star Wars vision could be as much as $240 billion over the next two decades, but that's the least of our problems. According to a *Los Angeles Times* account of a classified US intelligence assessment that was leaked to the press last May, deployment of an NMD [new missile defense] system by the United States is likely to provoke "an unsettling series of political and military ripple effects... that would include a sharp buildup of strategic and medium-range nuclear missiles by China, India and Pakistan and the further spread of military technology in the Middle East."[40]

Examination of four stages of U.S. foreign policy from 1946 to the present reveals that foreign policy is never static. There is an underlying consistency, however, with constant factors including "any means necessary" and the defense of corporate interests at the expense of the poor. The myth of corporate-led globalization is that it is good for almost everyone. Before looking at the role of the SOA/WHISC in the age of globalization (Chapter 9) I want to name globalization's high costs for majorities within and outside the United States.

Chapter 8

Globalization and Greed

The architects of U.S. foreign policy from 1946 to the present have been true believers. An "any means necessary" foreign policy is possible when advocates are convinced that the means they employ, whether the torturer's hand or the banker's rules, are justified because they promote the common good or protect particular interests they represent. Kennan said honestly (in once-secret documents) that U.S. foreign policy is about power politics and defending systems of inequality. He relegated concerns about democracy, development, and human rights to the realm of impossible altruism. The Hoover Commission did the same. It called on the United States to "subvert, sabotage, and destroy our enemies by more clever, sophisticated, more effective methods than those used against us."

National Security State ideology reflected these impulses. It was featured centrally in the discourse of U.S. leaders and their military allies throughout Latin America. According to this ideology, those who worked for social change were agents of international communism, considered evil, and warranted elimination. These values and views were expressed in CIA and SOA training manuals, and in the conduct of U.S.-trained SOA graduates and other military and paramilitary forces that murdered priests, nuns, bishops, campesinos, labor leaders, and students.

U.S. foreign policy embraces a philosophy of "any means necessary" because the people shaping and implementing it consider the tactics used to be necessary. This is true whether the tacticians are dissident generals carrying out coups, contras slitting throats, SOA graduates killing priests, or IMF workers implementing SAPs. Any means necessary is embraced also because *standing behind those tacti-*

cians are other strategists and decision-makers, including presidents who encourage Secretaries of State to arrange hits on generals committed to democracy in Chile; CIA directors to hire generals from Argentina to destabilize and destroy Nicaragua; trainers at the School of the Americas to teach low-intensity warfare strategy; and Secretaries of the Treasury and IMF directors to maximize economic leverage in pursuit of elite economic interests.

Most foreign policy actors, for reasons of public consumption, self-deception, or some combination of both, prefer to cast their policies and actions in light of noble missions. In his 1988 book, *Rethinking the Cold War,* journalist Eric Black noted that people need paradigms in order to "attach meaning to events," even when the meaning attached is false:

> Our paradigm is constantly reinforced by television shows, movies and plays, even by the vocabulary our leaders and our news media employ when they talk about the world. The United States has allies; the Soviet Union has puppets. Our side is run by governments, theirs by regimes. We have police, they have secret police. We engage in covert actions, they commit subversion. Our government puts out announcements, theirs propaganda....
>
> During the 1970s and 80s, the United States has supported governments in Latin America and elsewhere that rule by force, flout their own constitutions, rig elections, murder, torture, and oppress their own people. The State Department calls these countries "emerging democracies," [and] praises their improving human rights record.... The superpower that, according to our view of ourselves, respects international law, defied international law by mining the harbors of [Nicaragua], then refused to defend its actions before the World Court of Justice. The CIA created, armed, and trained the Contras ... to overturn the Nicaraguan revolution.... Our ideals and our conduct in the world are harder and harder to reconcile. Our euphemisms and self-deceptions are harder to accept. And yet ... the paradigm is alive and well.[1]

The New Mythology: Beneficial Globalization

A new mythology has emerged in the context of corporate-led globalization that builds on and replaces the old. It has three dimensions. First, globalization is inevitable. Globalization, Thomas Friedman writes, has "replaced the Cold War system as the dominant organizing framework for international affairs." "Globalization means the spread of free-market capitalism to virtually every country in the world. Globalization also has its own set of economic rules — rules that revolve around opening, deregulating and privatizing your economy." "Globalization isn't a choice," he writes. "It's a reality."[2]

The second dimension of the new mythology is that corporate-led globalization is not only inevitable but benefits most of us and is humanity's best and only hope for a prosperous future. Friedman's book names and discards negative consequences of globalization in favor of optimistic and glowing claims. In a world of unprecedented inequality, Friedman says, "Soon everyone will have a seat at the New York Stock Exchange."[3] In the context of urban slums that mar the human and geographic landscape in country after country, he writes of globalization's "own demographic pattern — a rapid acceleration of the movement of people from rural areas and agricultural lifestyles to urban areas and urban lifestyles more intimately linked with global fashion, food, markets and entertainment trends." And in a world in which poor people have been ravaged by structural adjustment programs and the needs of the Electronic Herd, Friedman speaks positively of a new role for mom and pop investors. "Suddenly [as a result of junk bonds] you and I and my Aunt Bev could buy a piece of Mexico's debt, Brazil's debt, or Argentina's debt — either directly or through our pension and mutual funds. And those bonds traded every day, with their value going up and down according to each country's economic performance." He then cites appreciatively Joel Korn, who headed Bank of America Brazil:

> After extending these loans, the banks, instead of just carrying them on their books, chopped these loans up into U.S. government-backed bonds that were sold to the public. That brought thousands of new players into the game. Instead of a

country just dealing with a committee of twenty major commercial banks, it suddenly found itself dealing with thousands of individual investors and mutual funds. This expanded the market, and it made it more liquid, but it also put a whole new kind of pressure on the countries. People were buying and selling their bonds every day, depending on how well they performed. This meant they were being graded on their performance every day. And a lot of the people doing the buying and grading were foreigners over whom Brazil, Mexico, or Argentina had no control.

"If a country didn't perform," Friedman continues, "the public bondholders would just sell that country's bonds, say goodbye, and put their money into bonds of a country that did perform."[4] To which we, like a character on Saturday Night Live, might say, "Now isn't that nice." Less sarcastic but equally biting is the critique of global development offered by theologian and environmentalist John Cobb:

> "Development" in Third World countries is typically accompanied by similar costs. Millions of women and children have been sold into prostitution or have adopted that life as their only option. Poor peasants in Latin America who have now become poor dwellers in favellas have lost the cultural and communal support that once gave some meaning to their lives. Around the world, tens of millions of people, mostly women, who once eked a living from the soil, now labor for pitiful pay, cut off from family and village life, sleeping in worker dormitories. By World Bank statistics, they are part of the success. They now earn more than a dollar a day, but for them, too, the human cost is enormous.[5]

The third dimension of the new paradigm recasts the myth of the benevolent superpower (a myth that masked murderous atrocities during the Cold War) in light of globalization. The old myth, Black notes, said that we "were Number One and the whole world loved us because all we wanted was peace, prosperity, freedom, and democracy for everyone."[6] The new myth says that the United States promotes globalization (never called corporate-led) as an act of benevolence. "Sure, it may seem unfair that America assumes a disproportionate burden for sustaining globalization," Friedman writes. "It means there are a lot of free riders...."[7] "America," Friedman writes, "truly is the ultimate benign hegemon and reluctant enforcer."[8]

From Myth to Reality

As we hear the new propaganda, we should remember that self-deception, personal or national, is always costly. This was true during the Cold War and it is true today. The new myth can be challenged with three counterclaims. First, globalization is a reality but corporate-led globalization is neither inevitable nor desirable. Second, far from benefiting most people and being humanity's best hope for a prosperous future, corporate-led globalization threatens to destroy life on earth as we know it. Third, the United States isn't a benevolent superpower.

Like many other nations the United States has shown itself capable of compassionate action, particularly in the context of international emergencies. U.S. foreign aid as a percentage of GNP, however, is pathetically low and much of the aid provided is military rather than developmental, or is given to strategic allies like Israel, Egypt, and Colombia. Groups such as Bread for the World and the Jubilee 2000 coalition have mobilized public pressure sufficient to achieve some gains in the quality of foreign aid and some reduction in the debt burden of the world's poorest countries. These important gains do not alter the general pattern reflected in U.S. foreign policy. The United States isn't and never has been a benevolent superpower or "benign hegemon." Its "any means necessary" foreign policy in service to elite economic interests has been in place throughout the four stages of U.S. foreign policy discussed in previous chapters.

The architects of globalization, like the generals who enforced the dictates of national security states, have strong convictions, a powerful ideology, and control of mechanisms of power. They have, as the previous chapter made clear, numerous options from which to choose as instruments of power projection. As Friedman notes:

In a winner-take-all world, America, for the moment at least, certainly has the winner-take-a-lot system. This makes America a unique superpower. It excels in the traditional sources of power. It has a large standing army, equipped with more aircraft carriers, advanced fighter jets, transport aircraft and nuclear weapons than ever, so that it can project more power farther than any country in the world. And deeper too. The fact that America has both a B-2 long-range stealth bomber and the short-range F-22 stealth fighter now being developed means that the U.S. Air Force can fly into any other country's air defense system virtually undetected. At the same time, as detailed above, America excels in all the new [economic] measures of power in the era of globalization.[9]

Undermining Democracy

Corporate-led foreign policy in the age of globalization, in sharp contrast to the new mythology of "the ultimate benign hegemon" and public promises of widespread benefits, can be as destructive to the poor as military repression and it threatens all of us. Its mechanisms of power undermine democracy, as the previous chapter demonstrated. One could argue that the constricted political democracy within the framework of economic life dictated by the IMF, the "Golden Straitjacket," the "Supermarkets," and the "Electronic Herd" is a step above third world dictatorship. It is also far from authentic democracy. Friedman, speaking to both developed and underdeveloped countries, says, "on the political front, the Golden Straitjacket narrows the political and economic policy choices of those in power to relatively tight parameters." Put on the Golden Straitjacket and "your economy grows and your politics shrinks." Its "political choices get reduced to Pepsi or Coke — to slight nuances of taste...."[10] The "Electronic Herd gets to vote in all kinds of countries everyday, but those countries don't get to vote on the Herd's behavior...."[11] As Wharton School globalization expert Stephen J. Kobrin writes, "when the power shifts to these transnational spheres, there are no elections and there is no one to vote for."[12]

Inequality

Democracy is not the only casualty. Corporate-led globalization aggravates problems of poverty and inequality, and its definition of life's meaning and its vision of the future are tragically inadequate to the requirements of the present crisis and future needs. Economist Xabier Gorostiaga says corporate driven globalization creates a "champagne glass" civilization. The richest 20 percent hoard 83 percent of the wealth while 60 percent of the world's people live and die on 6 percent. Huge numbers of people are considered "disposable" because they are not needed for production or consumption.[13] "Although 200 million people saw their incomes drop between 1965 and 1980, more than 1 billion people experienced a drop from 1980 to 1993."[14]

A U.N. Human Development Report states bluntly: "Global inequalities in income and living standards have reached grotesque proportions."[15] The "three richest people have assets that exceed the combined GDP [Gross Domestic Product] of the 48 least developed countries." The richest 225 people have combined incomes greater than those of half of humanity. Nearly 3 billion people struggle to live on less than $2 a day.

The situation is desperate but easy to improve with less greed and more political will to share the enormous bounty of the earth. Developing countries could achieve and maintain "universal access to basic education for all, basic health care for all, reproductive health care for all women, adequate food for all, and safe water and sanitation for all" at a cost of approximately 40 billion additional dollars a year. "This is less than 4 percent of the combined wealth of the 225 richest people in the world."[16]

The vast majority of U.S. citizens have reasons for concern. The U.S. is leading the charge to corporate-led globalization because a relatively small percentage of its citizens determine and benefit from policies that aggravate destructive inequalities at home and abroad. The U.S. is the most unequal of all industrial countries. One of four U.S. children is born into poverty. Approximately forty-seven million Americans have no health insurance. *The wealth of the richest*

1 percent of U.S. households is greater than the combined total of the bottom 95 percent. In 1998, Bill Gates, whose wealth is more than the bottom 45 percent of American households combined, increased his net worth by more than $2 million an hour.[17] In the United States today, "the top 1 percent of income-earners, 2.7 million people receive 50.4 percent of the national income, *more than the poorest 100 million people.*"[18]

As we have noted, Friedman presents a rosy portrait of mom and pop investors buying a piece of Mexico's debt and a romanticized view that soon "everyone will have a seat on the New York Stock Exchange." Such musings ignore the fact that a powerful minority determines the Golden Straitjacket, constitutes the Electronic Herd, and controls the Supermarkets. For example, "the wealthiest 10 percent of the population owns almost 90 percent of all U.S. business equity, 88.5 percent of bonds, and 89.3 percent of stocks."[19] Almost 90 percent of the increase in the stock market in the 1990s went to the top 10 percent of households, and 42 percent went to the richest 1 percent.[20]

The privileged few in all nations, those with surplus capital, participate in the Electronic Herd and benefit from imposition of the Golden Straitjacket. The vast majority of the U.S. and the world's people, however, struggle to make ends meet based on wages earned. Wage inequalities in the United States, like the wealth gap, are shocking. In 1975 the gap separating the average paid worker in a U.S. firm and the highest paid worker was 41 to 1. In 1998, it was 419 to 1.[21] It is 475 to 1 today. A minimum wage worker living in a home financed by Green Tree Financial would have to work more than *9,520 years* in order to earn as much as the CEO of Green Tree Financial earned in *one year.*[22]

Despite booming stock markets and talk of a robust economy "inequality in wages is at an all time high," and many U.S. workers are losing ground. Chuck Collins and Felice Yeskel of United for a Fair Economy write:

> During the last twenty years, three out of four U.S. wage earners have lost ground on the job. In real terms, this means that

people's wages have not kept up with inflation or that workers have lost some portion of benefits they previously had.[23]

Globalization is a significant cause of growing inequality, as Friedman acknowledges. During "the 1980s and 1990s, as globalization replaced the Cold War system, income gaps between the haves and have-nots within industrialized countries widened noticeably...."[24] Friedman cites economists Robert Frank and Philip Cook who state that globalization "has played an important role in the expansion of inequality" by creating a winner-take-all market for the globe.[25] "As Frank and Cook point out," Friedman writes, "while the winners can do incredibly well in this global market, those with only marginally inferior skills will often do much less well, and those with few or no skills will do very poorly. Therefore," he continues, "the gap between first place and second place grows larger, and the gap between first place and last place becomes staggering."[26] He notes that in 1982 there were 13 billionaires but by 1998 there were one hundred seventy.[27]

Friedman provides substantial support for critics who say that corporate-led globalization is a race to the bottom. The Supermarkets "play Syria off against Mexico off against Brazil off against Thailand. Those who perform are rewarded with investment capital from the Supermarkets. Those who don't are left as roadkill on the global investment highway."[28] The desire to please foreign investors and avoid becoming "roadkill" places governments at odds with their own workers. A *New York Times* article, "Labor Progress Clashes With Global Reality," introduces us to Abigail Martínez. Six years ago she made "55 cents an hour sewing cotton tops and khaki pants" for the Gap while working 18-hour days in an unventilated factory. Due to international pressure the factory is now ventilated, with unlocked bathrooms, and her workday is shorter. Still, Ms. Martínez earns only 60 cents an hour. "The lesson from Gap's experience in El Salvador," Leslie Kaufman and David Gonzalez write, "is that competing interests among factory owners, government officials, American managers, and middle-class consumers — all with their eyes on the lowest possible cost — make it difficult

to achieve even basic standards and even harder to maintain them."
They continue:

> El Salvador must compete with neighbors like Honduras and
> Nicaragua, where wages are lower and the population even
> poorer and more eager for work. Government officials and fac-
> tory managers concede that El Salvador's current minimum
> wage is not enough to live on — by some estimates it covers
> less than half of the basic needs of a family of four — but
> they are wary of increasing it. "We cannot be satisfied with
> the wage, but we have to acknowledge the economic realities,"
> Mr. Nieto [labor minister] said.[29]

If you don't want to be "roadkill on the global investment highway" then you must pay "homekill" wages. Several additional quotes
from Friedman demonstrate that corporate-led globalization is a race
to the bottom both in terms of environmental and labor standards:

> A commodity is any good, service, or process that can be pro-
> duced by any number of firms, and the only distinguishing
> feature between these firms is who can do it the cheapest
> (p. 66).
>   Increasingly, there was a single, open global marketplace, and
> Cyberspace, where a multinational company could sell anything
> anywhere or make anything anywhere. This has sharpened
> competition and squeezed profit margins in many industries.
> As a result, every big multinational needs to try to sell globally,
> in order to make up in volume for shrinking profit margins,
> and it needs to try to produce globally — by slicing up its pro-
> duction chain and outsourcing each segment to the country
> that can do it the cheapest and most efficiently — in order to
> keep manufacturing costs down and remain competitive. This
> has led to more multinationals investing in more cost-lowering
> production facilities abroad, or making alliances with cheaper
> subcontractors abroad — not to survive in a world of walls,
> *but to survive in a world without walls* (p. 111, emphasis in
> original).

A lot of the foreign investing that the long-term cattle [investors who don't move investments in and out daily or hourly] do these days is not building factories anymore. It is developing alliances with locally owned factories, which serve as affiliates, subcontractors, and partners of the multinational firms, and these production relationships can be and are moved around from country to country, producer to producer, with increasing velocity in search of the best tax deals and most efficient and low-cost labor forces. The long-term cattle play off every developing country against the other. Each of these countries is desperate for multinational investments.... Nike first established its Asian production facilities in Japan, but when that got too expensive it hopped over to Korea and then went to Thailand, China, the Philippines, Indonesia and Vietnam (p. 112).

India is rapidly becoming the back office for the world. Swissair moved its entire accounting division, including computers, from Switzerland to India to take advantage of lower labor costs for secretaries, programmers, and accountants. Thanks to digitization and networking Swissair can have its bookkeepers in Bombay today as easily as in Bern (p. 46).

The "roadkill on the global investment highway" includes workers in the United States who are forced into competition with low-wage workers worldwide. Researchers at United for a Fair Economy describe "free trade" as a weapon against U.S. labor:

The North American Free Trade Agreement (NAFTA) and the specter of jobs moving to Mexico has led to a pervasive sense of job insecurity among American workers, undermining union bargaining power and weakening wage demands.

The results of a study of the effects of plant closings or threat of plant closings on the right of workers to organize in the United States, conducted by Kate Bronfenbrenner, Cornell University, show that plant closing threats are an extremely pervasive and effective component of employer anti-union strategies. Employers threaten to close the plant in 50% of all union election campaigns and 52% of all withdrawals. Plant closing threats also have a devastating impact on union win rates, with unions only able to win 33% of the campaigns where plant closing threats occurred, compared to 47% win rate in campaigns where no threats occurred.[30]

NAFTA has been hard on Mexican workers too. A detailed study, "A Hemisphere for Sale: The Epidemic of Unfair Trade in the Americas," includes this assessment:

> The results of a trade agreement between a developing country and the world's largest economy are as anyone would predict. Seven years after NAFTA's implementation, U.S. corporate profits are skyrocketing. In fact, the economic elite on both sides of the border reap huge benefits, while the majority of Mexicans watch their buying power drop and wages stagnate.
>
> Once trumpeted as Mexico's gateway into the developed world, NAFTA has done little to bring true development to Mexico. Many Mexicans ... — who were not consulted or involved in its planning — have been harmed by the agreement. NAFTA has not fulfilled its promise of decreasing poverty: since 1994, poverty in Mexico has increased from 66 percent to 70 percent. Nearly half of Mexicans now earn less than three dollars a day. The wealth generated by NAFTA has concentrated dramatically in a few hands; Mexico has more billionaires than any other developing nation, while most workers earn less than one percent more than they did 18 years ago.[31]

Free trade agreements and corporate-led globalization are devastating to agriculture, as the prior discussion of Colombia demonstrated. I am speaking not only of the environmental impacts — chemical fertilizer use and water contamination, pesticide use, genetically modified foods, or energy wasted when food travels thousands of miles to market, but of the profound economic and social costs that accompany disruption of rural life, the erosion of community, the loss of livelihood. Friedman talks about agriculture in this way:

"Either you got big and were able to take advantage of economies of scale and played in the global farmers market, or you got swallowed up by someone else who could." Talking to a successful U.S. farmer in the age of globalization, he writes about the "key strategy which is making his farm bigger so it will be the one that eats others and is not eaten itself."[32]

The promised benefits of the global farmers market have reached few U.S. or third world farmers. Many have been eaten by privileged survivors who themselves get eaten until you have happy marketers of grains and other food stuffs together with rural crises in developed and underdeveloped countries alike. Witness for Peace describes the impact of NAFTA on one Mexican farmer:

> Ten years ago, Javier Pérez could provide for his family. He grew enough corn and beans on his small plot of land in southern Mexico to feed his wife and five children, and sold his extra harvest for money to buy shoes, schoolbooks, and other necessities. Over the years, Javier earned enough to send his five children to primary school, and fix up his modest, dirt-floor house.
>
> About five years ago, Javier lost his markets for corn and beans. Upon Mexico's entry into the North American Free Trade Agreement (NAFTA) with Canada and the United States in 1994, imported U.S. corn and other basic foods flooded the Mexican market, leaving Javier with nowhere to sell his crops. At the same time, cuts in government support to small farmers raised his cost of production.
>
> In recent years, Javier has planted papaya, cantaloupe, tomatoes, and watermelon in turn. But without resources, technology, or a secure market, he ended up with a barn full of rotting fruit and a growing debt with the bank. Like the majority of Mexicans, Javier is now poorer than his parents were. In order to keep the family afloat financially, his eldest son has already left for the United States, where he found work as a migrant farm worker and sends money home periodically. Another son and daughter are considering emigrating.[33]

Friedman and other advocates are fully aware that corporate-led globalization feeds inequalities. They say their version of globalization is a fact of life and we have no choices. If this is true then we are all doomed. What they don't say is that corporate-led globalization benefits perhaps 20 percent of the world's people and is a bust for most of the rest of us. Paul Street, research director at the Chicago Urban League, writes:

> Beneath the chorus [of voices proclaiming the benefits of globalization], those who cared to listen could discern from within the mainstream media discordant notes of deepening human suffering and shocking inequality. . . . In the candid words of the *Boston Globe*, "globalization" had "resulted in a boom for the wealthiest 20 percent of the world's population and a bust for just about everyone else." . . . Correspondent RC Longworth of the *Chicago Tribune* marked the millennium's turn by noting that the world's "surging economy enriches a few" but "bypasses the rest." . . . Those lucky people were a distinct minority for whom the new global era was "a golden age of peace, great wealth, booming markets. Easy travel. Instant communications, fabulous comfort and, with it, an innocence and confidence that this good fortune is not only deserved but permanent." But "things are very different," Longworth noted, for the world's "majority [who] . . . live in shanty towns on the outskirts of the global village." Longworth referred to "the rest of humanity" beneath the opulent minority: "millions of unemployed nomads in China, street people in Calcutta, European workers without jobs, the 28 percent of Americans whose jobs pay poverty-level wages, semi-educated young men in Morocco begging in four languages, the hopeless poor of Africa, child laborers in Bangladesh, the pensioners of Poland, the Russians wondering what happened to their lives."[34]

Inequality is not a bad thing for the lucky 20 percent, including Friedman who drives a Lexus and uses the Lexus as a positive symbol of globalization. "Unless you're a Lexus car dealer or a diamond jeweler," Collins and Yeskel write, "inequality is a threat to wider prosperity in our economy."[35]

The Environment

Inequality and other aspects of corporate-led globalization also threaten the environment. Businessman Paul Hawken writes:

> Quite simply, our business practices are destroying the life on earth. Given current corporate practices, not one wildlife reserve, wilderness, or indigenous culture will survive the global market economy. We know that every natural system on the planet is disintegrating. The land, water, air, and sea have been functionally transformed from life-supporting systems into repositories for waste. There is no polite way to say that business is destroying the world.... How do we imagine a future when our commercial systems conflict with everything nature teaches us?
>
> Whatever possibilities business once represented, whatever dreams and glories corporate success once offered, the time has come to acknowledge that business as we know it is over. Over because it failed in one critical and thoughtless way: It did not honor the myriad forms of life that secure and connect its own breath and skin and heart to the breath and skin and heart of the earth.[36]

Development expert David C. Korten agrees that corporate-led globalization involves corporate-driven environmental destruction. "Increasingly, it is the corporate interest more than the human interest," he writes, "that defines the policy agendas of states and international bodies."[37] A study by the Institute for Policy Studies indicates that of the 100 largest economies in the world, 51 are corporations and 49 are countries. The top 200 corporations combined are the equivalent of 27.5 percent of the world's economic activity but they employ only 0.78 percent of the world's workforce.[38] We "are now coming to see that economic globalization has come at a heavy price." Korten notes that we are placing "human civilization and even the survival of our species at risk mainly to allow a million or so people to accumulate money beyond any conceivable need."[39]

Worldwatch Institute describes key environmental issues through its magazine (*World Watch*) and its annual assessment (*State of the World*). A sense of urgency pervades its documents. "If the world is to achieve sustainability," *State of the World 1991* states, "it will need to do so within the next 40 years. If we have not succeeded by then, environmental deterioration and economic decline are likely to be feeding on each other, pulling us into a downward spiral of social disintegration."[40] The March–April 2001 issue of *World Watch* magazine warns "that the rate of environmental destruction has reached warp speed."[41] Several key environmental issues addressed consistently by Worldwatch demonstrate that in relation to the most critical environmental problems facing the world and the human family today corporate-led globalization and a corporate-driven U.S. foreign policy aggravate problems and block solutions.

First, ending poverty is an environmental necessity. At present the rich not only do disproportionate harm to the global commons; they also force the poor to degrade their environment in order to survive. Poor countries have often destroyed tropical forests and degraded other land and water resources in response to IMF conditions and trade agreements requiring them to export more and open their economies to foreign penetration. Rich landowners positioned to benefit from these agreements push landless peasants onto marginal lands, where they deplete fragile soils and clear-cut forests. "Poverty's profile . . . has become increasingly environmental," Alan Durning writes. "The poor not only suffer disproportionately from environmental damage caused by those better off, they have become a major cause of ecological decline themselves as they have been pushed onto marginal land by population growth and inequitable development patterns."[42]

Resulting environmental degradation threatens rich and poor alike. The quality of air we breathe in North America is connected to tropical forests in the Amazon. Ending poverty, therefore, has become an environmental as well as moral imperative. As Sandra Postel of Worldwatch writes, "the future of both rich and poor alike hinges on reducing poverty and thereby eliminating this driving force of global environmental decline."[43] Reducing poverty is also an important factor in reducing population growth, a serious global problem that receives strikingly little attention from U.S. leaders, who under pressure from religious conservatives, refuse to fund adequately programs of the UN population agency.[44] Corporate-led globalization, as detailed above, concentrates wealth and fosters inequality. An any means necessary foreign policy, whether a war against the poor using bullets or bankers, reinforces systems that reward the privileged few at the expense of the many, leading all of us to the edge of environmental collapse.

A second environmental imperative is a rapid transition away from fossil fuels and nuclear power to renewable energy resources. "Solar energy will be the foundation of a sustainable world energy system," according to Worldwatch.[45] There is some good news. Wind energy has expanded dramatically in recent years. President George W. Bush, however, is promising to revitalize nuclear power and to make increased production of fossil fuels the linchpin of U.S. energy policy. Excessive fossil fuel use and carbon emissions are the result of unsustainable products and lifestyles. They contribute to global warming that threatens to undermine food production and to place wetlands, coastal forests, and cities at risk as sea levels rise.

The Intergovernmental Panel on Climate Change (IPCC) established by the United Nations and the World Meteorological Organization (WMO) in a 2001 report notes that "the 1990s was the warmest decade and 1998 the warmest year" since 1861 and that "data for the Northern Hemisphere indicate that the increase in temperature in the 20th century is likely... to have been the largest of any century in the past 1000 years." "There is," according to the IPCC, "new and stronger evidence that most of the warming observed over the last 50 years is attributable to human activities" and that the degree and consequences of global warming are worse than previously thought. These human-induced changes "will persist for many centuries."[46]

The U.S. unilateral withdrawal from the Kyoto agreement to curb carbon emissions, support for repressive Gulf states, plans to drill for oil in pristine wilderness preserves in Alaska, cuts in funds for conservation programs and alternative energy development, plans to revitalize the nuclear industry, billions of Pentagon dollars protecting oil supplies in the Middle East, and counterinsurgency warfare in oil-rich Colombia all reflect powerful corporate agendas at the heart of the global system and U.S. foreign policy today. These agendas clash sharply with efforts to avoid a series of environmental catastrophes.

A third critical environmental issue is the need to move away from present patterns of production and consumption and redefine the meaning of life. Corporate-led globalization contributes to global warming because it globalizes unsustainable lifestyles symbolized by gas-guzzling SUVs and luxury vehicles like Friedman's prized Lexus. Following World War II, retailing analyst Victor Lebow laid out the foundational values of the market economy:

> Our enormously productive economy... demands that we make consumption our way of life, that we convert the buying and use of goods into rituals, that we seek our spiritual satisfaction, our ego satisfaction, in consumption.... We need things consumed, burned up, worn out, replaced, and discarded at an ever increasing rate.[47]

Corporate-led globalization extends these values to every corner of the globe. "Today, for better or for worse," Friedman writes, "globalization is a means for spreading the fantasy of America around the world."[48] "Culturally speaking, globalization is largely, though not entirely, the spread of Americanization — from Big Macs to iMacs to Mickey Mouse — on a global scale."[49] This spread of a "homogenizing" dominant culture is potentially disastrous, as Friedman himself recognizes:

> Because globalization is creating a single marketplace — with huge economies of scale that reward doing business or selling the same product all over the world all at once — it can homogenize consumption simultaneously all over the world. And because globalization as a culturally homogenizing and environment-devouring force is coming on so fast, there is a

real danger that in just a few decades it could wipe out the ecological and cultural diversity that took millions of years of human and biological evolution to produce.[50]

Wasteful U.S. lifestyles are themselves unsustainable. Spreading them globally accelerates all major environmental stresses. The problem is not only one of resource depletion or pollution, however; it is also familial, cultural, and communal. "Mutual dependence for day-to-day sustenance ... bonds people as proximity never can," Alan Durning writes. "Yet those bonds have severed with the sweeping advance of the commercial mass market...." Durning continues:

> Like the household, the community economy has atrophied —
> or been dismembered — under the blind force of the money
> economy. Shopping malls, superhighways, and "strips" have
> replaced corner stores, local restaurants, and neighborhood the-
> aters — the things that help create a sense of common identity
> and community in an area. Traditional communities are all but
> extinct in some nations.... The transformation of retailing is
> a leading cause of the decline of traditional community in the
> global consumer society.[51]

The corporate-led global economy can deliver an overabundance of goods to a minority of people across national boundaries but it can't deliver essential goods to all or a sense of meaning, community, identity, or purpose to most. As David C. Korten writes:

> The leaders and institutions that promised a golden age are
> not delivering. They assail us with visions of wondrous new
> technological gadgets, such as airplane seats with individual
> television monitors, and an information highway that will make
> it possible to fax messages while we sun ourselves on the beach.
> Yet the things that most of us really want — a secure means of
> livelihood, a decent place to live, healthy and uncontaminated
> food to eat, good education and health care for our children, a
> clean and vital natural environment — seem to slip further from
> the grasp of most of the world's people with each passing day.
> Fewer and fewer people believe that they face a secure economic

future. Family and community units and the security they once provided are disintegrating. The natural environment on which we depend for our material needs is under deepening stress.[52]

As James Surowiecki notes, the rapid pace of change within corporate-led globalization "makes the system a terrific place for innovation, [but] it makes it a difficult place to live, since most people prefer some measure of security about the future to a life lived in almost constant uncertainty. . . ."[53]

Finally, the transition to a sustainable world depends on demilitarization, new definitions of security, alternatives to war, and massive conversion of materials and human talent away from weapons production into active efforts on behalf of peace and environmental integrity. "Despite growing public concern," *State of the World 1990* says, "government expenditures to defend against military threats still dwarf those to protect us from environmental ones. For example, the United States plans to spend $303 billion in 1990 to protect the country from military threats but only $14 billion to protect from environmental threats, a ratio of 22 to 1."[54] More than a decade later this ratio is widening.

It is in relation to militarization that the claims of corporate-led globalization advocates diverge most sharply with reality. I mentioned trends in the previous chapter, including a temporary downsizing of the role of U.S. and Latin American military forces, a resurgence of militarization in Stage 4, human rights talk in the context of constricted democracy, and ongoing repression in Colombia. Friedman reminds those who think economic power can substitute completely for military power that "globalization does not end geopolitics." He notes that countries that have McDonald's don't go to war due to economic integration and U.S. military power and warns that the "hidden hand of the market will never work without the hidden fist. McDonald's cannot flourish without McDonnell Douglas, the designer of the U.S. Air Force F-15." Most important, Friedman knows that although elites in many nations are less likely to go to war with each other because in a globalized world they have common interests, *they will frequently be at war with their own people*. Stages 3 and 4 in U.S. foreign policy inevitably lead back in the direction of the repression of Stages 1 and 2. Friedman says it this way:

> The globalization era may well turn out to be the great age of civil wars. In these new civil wars, the battle lines will not be between pro-Americans and pro-Soviets, or even between traditional left and traditional right. No, these civil wars will be between pro-globalizers and anti-globalizers, between globalists in each society and localists in each society, between those who benefit from change and from this new system and those who feel left behind by it.... That's why, when people ask me what I do for a living, I sometimes answer, "I'm the foreign affairs columnist for the *New York Times* and I cover the wars between winners and losers *within* countries."[55]

It is in this context that we can understand the current role of U.S. foreign policy, including the SOA/WHISC.

Chapter 9

A Rose by Any Other Name

You would be upset if the United States helped serial killer Charles Manson, from your neighborhood and without any treatment or expression of remorse, legally change his name to Nice Fairchild and return to his previous residence. You should be equally upset that the Pentagon with support from the White House treats the long record of abuses linked to the School of the Americas as a public relations problem. The latest Pentagon-led fiasco concerning the school involves the school's name change to the Western Hemisphere Institute for Security Cooperation (WHISC).

Behind the Name Change

The impetus for the name change comes from the success of the movement to close the SOA.[1] The U.S. House of Representatives voted 230–197 in 1999 to prohibit use of Foreign Operations funding for the U.S. Army School of the Americas (SOA). This vote demonstrated that more than a decade of citizen protest, research, public education, lobbying, and civil disobedience was bearing fruit. Disgust with a school of assassins linked to human rights atrocities throughout Latin America had reached the broader public and a majority of House members. The vote to cut funds for the SOA was thwarted in conference committee with the Senate. SOA defender Joseph C. Leuer describes what happened:

> The fight had now been pushed to a conference committee with the Senate, which, according to key staffers, was hoping to avoid having to table the issue by keeping such a raucous

119

debate in the House of Representatives. At the behest of the Secretary of the Army, Col. Weidner visited key members to personally brief them on USARSA's accomplishments. A joint committee later voted 8–7 against passing to the Senate a bill that deleted funding USARSA for fiscal year 2000. This was a small victory, but one that signaled future compromises if the U.S. Army wanted to continue with a school dedicated to the professional education and training of our Latin American allies.[2]

The House vote to cut funding, though unsuccessful, got the attention of SOA supporters within the school, the Pentagon, and the White House whose objectives and counterstrategy were simple. They wanted to deflate the growing movement to close the SOA, get the horrific abuses of the SOA and its graduates out of the public spotlight, and allow the school to continue its foreign policy mission. To achieve these objectives they engaged in intensive lobbying of members of the House and Senate. As a result, both houses of Congress voted on the same day in December 2000 to close and reopen the SOA in January 2001 as the Western Hemisphere Institute for Security Cooperation (WHISC).

Closing the SOA and reopening it at the same site with essentially the same curriculum was a creative act of desperation. Faced with overwhelming evidence linking the SOA and its graduates to terror, torture, dictatorships, and lawlessness, school supporters responded for several years with a strategy of denial, cosmetic change, and damage control. These tactics failed. Colonel Mark Morgan told Congressional aides at a Defense Department briefing just prior to a critical vote: "Some of your bosses have told us that they can't support anything with the name 'School of the Americas' on it. Our proposal addresses this concern. It changes the name."[3] Paul Coverdell, the late Georgia Senator and SOA supporter, told the *Columbus-Ledger Enquirer* that with the name change the "School of the Americas would still be able to continue its purpose" and that the proposed changes to the school were "basically cosmetic."[4] In a December 2000 interview in *El Tiempo,* Colombian Defense Minister Luis Fernando Ramírez and Commander of the Armed Forces General Fernando Tapias stated that Congress and the U.S. government had assured them that the School of the Americas would continue to function following the name change and that members of the Colombian military could still train there.[5]

Many in Congress were not fooled by the name change that was approved by only a few votes in the House. The late Representative Joseph Moakley of Massachusetts called the name change the equivalent of "pouring perfume on a toxic dump." Today, there are bills in both the Senate and House to close the SOA/WHISC and the movement to close the school is strong.[6]

An editorial in the *National Catholic Reporter* called the name change "Orwellian." It went on to say:

> Except for denying the easy-to-manage sound of SOA, the changes do nothing to deal with the far deeper underlying problem. For the School of the Americas is more than a collection of courses and some errant foreign students. The school is symbolic of both a state of mind of the military forces and the training that goes on in many more places than Fort Benning, Ga.[7]

Further, it is not training done in isolation. The commando tactics and other techniques and methods of intimidation taught at the school may be carried out by soldiers of other countries, but they are carried out in service of U.S. policy. Ultimately, then, the gruesome record of SOA graduates is one that we, as a people, sanction because they act ostensibly in our interest. Surely we wouldn't continue to train them if that were not the case. That is why a simple name change and cosmetic alterations won't do.

The SOA/WHISC is a window through which we glimpse the heart of U.S. foreign policy. The name change is an effort to brick over the window. Pentagon officials consider the movement to close the SOA dangerous because it sheds light on the repressive tactics at the heart of U.S. foreign policy. SOA commandant Colonel Weidner writes that the "School of the Americas is closing, having *accomplished its Cold War mission.*"[8] What Weidner doesn't say is that the mission was bloody and the tactics used in carrying it out violated all legal and moral standards. The new director of the SOA/WHISC, Col. Richard D. Downie, confirms that the school has much to hide. Defending WHISC in an interview in the *Columbus Ledger Enquirer,* he says: "I was charged with taking a look at what was needed — *not necessarily at what the School of the Americas did or why it did those things,* because the School of the Americas was designed for the twentieth century Cold War paradigm."[9]

When Congress "closed" the SOA and "opened" the WHISC at the same site with essentially the same curriculum its action reflected more than cynicism. Members of Congress were under intense pressure. The Pentagon and the White House want desperately to place a permanent shutter over the SOA window because it provides too clear a view of U.S. foreign policy. SOA critic and former SOA instructor U.S. Army Major Joseph Blair says WHISC stands for "We Hide In a Semantic Cover-up of SOA's failures by sweeping a dirty past under a U.S. Army rug."[10]

Ed Kinane, an SOA Watch Board Member who served six months in prison for "crossing the line" onto the base at Fort Benning, echoed this sentiment in an article written for *The Post-Standard* newspaper in Syracuse, New York:

> *Despite* the facelift, the SOA/WHISC hasn't repudiated its ways, hasn't confessed its sins. There's been no Truth Commission. Neither a grand jury nor a war-crimes tribunal has been convened to explore indicting its commanding officers, nor have they yet served time. The survivors of the myriad victims have never received reparation. The Pentagon has fought tooth and nail to save the SOA, both in name and substance. For the Pentagon, civilian-initiated abolition equals loss of face. Its Latin American clients would be bewildered that U.S. civilians could possibly win a campaign to close the SOA. After all, isn't the SOA really about controlling civilians, stifling their dissent? Isn't anti-civilian warfare its main thrust?[11]

Using a tone of barely concealed rage Joseph Leuer described in his official history of the SOA how the school was nearly shut down because it was "an institution bloodied before public opinion by leftist theatrics, Congressional gamesmanship, and half-hearted support for USARA's mission from higher headquarters." It was only through "high level Inter-Agency contacts ... that the school was able to retaliate using its own internal assets...." "If these relationships had not been established," Leuer continues, "the school would have long since closed, unfairly tarnishing the U.S. Army and all the soldiers and civilians at the school who had performed their duties lawfully and faithfully since 1949." According to Leuer:

> The political fighting had deeply distressed the staff and faculty of the SOA. Over 200 loyal American soldiers and civilians who were dedicating the best of their service to improving militaries of the region for good ends felt they had been systematically defamed and seemingly abandoned by their senior leaders.[12]

This is not language telling the story of the SOA's closing. Leuer is describing how supporters mobilized "high-level Inter-Agency contacts," to keep the school open, give it a name change, and continue its mission. Why keep it open? What is the school's present mission in the context of corporate-led globalization?

The SOA/WHISC Mission Today

The similarities and differences that mark the four stages of U.S. foreign policy discussed earlier make it clear that people and agencies responsible for foreign policy learned an important lesson. Repressive militaries can be useful in one setting and unnecessary or counterproductive in another. The key challenge facing U.S. foreign policy planners and the SOA/WHISC today is to know the difference and implement policies accordingly. At the heart of foreign policy in Stage 4 is a powerful, disturbing contradiction. In the age of corporate-led globalization the preferred means of power projection are economic but the policies and practices of the IMF, NAFTA, the WTO, the Golden Straitjacket and the Electronic Herd are destabilizing. Conditions favorable to the corporate architects of the global system can often be imposed with instruments of economic leverage, but these very conditions aggravate the causes of social turmoil. Poverty, inequality, cultural uprooting, environmental decay, the undermining of local agriculture, and political frustration are the fruits of the corporate-led system. People denied a meaningful voice in shaping political and economic affairs that impact their lives, communities, and nations invariably deny corporations and U.S. leaders the stability they prize. Disenfranchised people organize and resist. As Friedman notes, if participation in the global system "comes at the price of a country's identity, if individuals feel their olive tree roots crushed, or washed out, by this global system, those olive tree roots will rebel. They rise up and strangle the process";[13] and that "in many countries, instead of popular mass opposition to globalization, is wave after wave of crime — people just grabbing what they need. . . . "[14] In either case, rebellion or crime, people must be stopped. "The globalization system cannot hold together," Friedman writes, "without an activist and generous American foreign policy." "The hidden hand of the market will never work without a hidden fist."[15]

The SOA/WHISC remains open today because it houses two key aspects of U.S. foreign policy under one roof: a military fist dimension and an economic fist dimension. The military fist is carried out through SOA/WHISC and other U.S. military training programs. Such training, both on and off the record of the official curriculum, equips Latin American soldiers to fight "civil wars," those described by Friedman as "wars between winners and losers *within* countries," between the beneficiaries and victims of corporate-led globalization.

Soldiers trained in military fist tactics are given the tools they need to carry out the traditional task of defeating movements, protestors, and others who organize alternatives to present injustices. These injustices include vast economic inequalities and constricted democracy, both predictable products and desired outcomes of corporate-led globalization. Historically and today the SOA/WHISC is associated with a foreign policy that legitimates and utilizes whatever military and paramilitary means are needed in any given setting, including tactics of terror and torture. The traditional tactics are most evident today in Colombia, the new El Salvador.

The second aspect of U.S. foreign policy housed at the SOA/
WHISC has to do with the economic fist dimension of U.S. foreign policy. The present SOA/WHISC is adapting to new needs
and opportunities that arise in the present geopolitical context of
corporate-led globalization in which economic leverage is the preferred, but not only, means of power projection. The United States
over many decades equipped the militaries of Latin America with
ideologies, weapons, and strategies to justify and carry out repressive
tactics. These militaries need a good deal of *additional training* in
the new geopolitical context. Having learned strategies of repression
requiring all the subtlety of a sledgehammer, they must now learn the
art of repression in the context of constricted democracy. They are
being schooled at the SOA/WHISC in the fine points of economic
diplomacy, psychological warfare, democracy enhancement, combat
skills, human rights, as well as repression. They are honing discernment skills in order to judge when and where repressive tactics are
still necessary, such as Colombia, and when and where economic
leverage is sufficient to achieve desired goals.

In El Salvador, Guatemala, and many other Latin American countries today, for example, human rights (political not economic) are
respected within the framework of constricted democracy. People can
vote for a wider range of candidates without getting killed, but the
Golden Straitjacket undermines the meaning of their vote. Friedman quotes democracy scholar Larry Diamond: "We have now seen
a number of examples where countries in Latin America, Eastern
Europe, and East Asia have voted out of office governments they
associated with the pain of globalization reforms. The new governments that came in," Diamond notes, "made some adjustments
but kept more or less the same globalizing, marketizing policies."[16]
"For the first time," Friedman writes, "virtually every country in the
world had the same basic hardware — free market capitalism."[17] "The
Electronic Herd turns the whole world into a parliamentary system,
in which every government lives under the fear of a no-confidence
vote from the Herd."[18] In this environment of constricted democracy "human rights" are to be respected and repressive militaries are
downsized, perhaps only temporarily.

The present role of the SOA/WHISC is determined within a geopolitical context in which third world economic debt, dependency, and vulnerability, and in some cases war-weary populations exhausted after decades of U.S.-sponsored repressive military violence, offer U.S. foreign policy planners many options of power projection. Leverage applied through the IMF, NAFTA-type agreements, the WTO, the Golden Straitjacket and the Electronic Herd are often more effective and less controversial than dictatorships and death squads.

The SOA/WHISC plays two important roles in the context of this economic fist aspect of U.S. foreign policy. One important mission is to give soldiers the military training they need to defeat enemies in settings where economic power projection is inappropriate, where it fails, or where "civil wars," civil disturbances, or other problems arise from the skewed and destructive policies of the IMF, NAFTA, the WTO, the Golden Straitjacket, and the Electronic Herd.

It is impossible to know if repressive military training remains the SOA/WHISC's most important mission, as it has been throughout the school's history. The United States trains Latin American soldiers in so many places under so many programs and pretenses that it is nearly impossible to monitor. What is most important for critics of the school to recognize is that the "any means necessary" philosophy and tactics are still in place within the school itself and within the broader spectrum of U.S. foreign policy. Many SOA graduates are involved in human rights atrocities in Colombia today. Colombian military leaders needed assurance that the name change didn't change training for their officers and soldiers. Most Colombian soldiers receive training, not at the SOA/WHISC, but in Colombia under the direction of U.S. special operations forces. These facts and trends testify both to the importance of the school and the flexible options available to U.S. leaders.

A second important mission today is that the SOA/WHISC serves as a major propaganda center on behalf of those who present the school, U.S. foreign policy, and corporate-driven globalization as instruments of freedom, democracy, and human rights. The policies

and practices of the economic fist of U.S. foreign policy are justified with a new mythology of benevolence, inevitability, and claims
of widespread benefits. The new mythology like the old covers up
important problems. Problems linked to corporate-led globalization,
including deadly inequalities, widespread environmental destruction,
and constricted democracy get smoothed over with an abundance
of democracy and human rights rhetoric spewing from the SOA/
WHISC. Colonel Patricio Haro Ayerve, Assistant Commandant of
the SOA, reflects this rhetorical spirit in an article written at the
time of the school's name change:

> After devoting 54 years to hemispheric security and the pro-
> fessional development of the members of the Armed Forces
> of all the countries of the Americas in an effort to ensure
> peace and democracy, the United States Army School of the
> Americas is closing its doors to usher in the Western Hemi-
> sphere Institute for Security Cooperation... Inter-American
> cooperation in security matters is fundamental, and, there-
> fore, it becomes necessary to establish a new institute with a
> continental focus; consequently, the School of the Americas,
> with complete satisfaction for having accomplished its mission,
> will "pass the torch" to the Western Hemisphere Institute for
> Security Cooperation.
>
> As Assistant Commandant of the School, it is my privilege
> to serve as the senior Latin American officer among the contin-
> gent of guest instructors.... Observing the ideals and principles
> that motivate them as they perform their professional duties in
> the American forum, let me therefore affirm that this faculty
> will continue wholeheartedly to exhibit the same enthusiasm
> and professional dedication to duty as they serve the new In-
> stitute. They will continue to fulfill their responsibilities with
> the same fervor that they have exhibited while assigned to our
> beloved School of the Americas.
>
> The School of the Americas has planted some very sig-
> nificant "seeds" in the "fertile soil" of what will soon be the
> Western Hemisphere Institute for Security Cooperation. These

elements include a heartfelt spirit of Americanism, present in the soldiers at the School, the profound conviction that all security activities should be conducted in accordance with international standards and with an absolute respect for human rights of individuals in conflicts, and the reaffirmation, embedded within each Latin American trained at the School, that unconditional subordination to the legitimately elected civilian government and the democratic civilian government system itself is the only form of government that involves genuine participation of the population in the transcendental governmental decisions regarding the well-being and progress of the citizenry. . . .

I would like to express my gratitude to the School of the Americas for having served more than half a century as a symbol of hemispheric brotherhood and the propagator of insightful teachings in security, democracy, and respect for human rights, despite beliefs to the contrary by the School's detractors.[19]

These are lofty claims for an institution known throughout Latin America as a school of dictators, assassins, and coups. Col. Haro Ayerve makes no effort to conceal the fact that the WHISC and the SOA are the same institution. The only real changes are that the "School's detractors" have forced a name change and that differences in the geopolitical context offer the school new challenges and opportunities. "The environment in which the School of the Americas was established has gradually changed over the years," Col. Haro Ayerve writes, "as the post Cold War era and *unipolarity* have produced a much different scenario, which, consequently, explains why the countries of the world have been forced to redefine and redirect their security issues."[20]

Unipolarity, as described through the lens of globalization advocate Thomas L. Friedman, refers not only to the fact that the United States is the only military superpower but that other nations have almost no options when it comes to whether or how to participate in the corporate-driven globalized economy dominated by the United States. The SOA/WHISC's effusive rhetoric about respecting human rights and democracy and the relatively recent addition and maintenance of courses on these topics in the SOA/WHISC curriculum need to be understood as a response to both the "School's detractors" *and* to the elevation of the economic fist dimension of U.S. foreign policy. Human rights rhetoric and courses, in other words, have something to do with both SOA/WHISC propaganda and additional roles assigned to the school in the changing geopolitical context.

Let's start with propaganda. A 1995 *Los Angeles Times* editorial calling for closing the SOA stated accurately that "it is hard to think of a coup or human rights outrage that has occurred [in Latin America] in the past 40 years in which alumni of the School of the Americas were *not* involved."[21] The link between the SOA, its graduates, and human rights abuses is so ironclad that it posed serious problems for SOA supporters. They responded by denying wrongdoing, with vague references to the Cold War, and by adding a number of courses on human rights and democracy to the curriculum. They then argued that these courses, not added until the mid-1990s, should be the lens through which people view the school's mission since its inception in 1946.

This classic propaganda is revealed in an information paper put out by the SOA in 1996. We are told that because of the SOA and "active U.S. military engagement . . . Latin America is today the least militarized and least violent region in the world." "All courses," the paper argues, "provide a forum in which military and civilian personnel from the entire Latin American region involved in defense-related issues come together to exchange ideas and create common bonds and respect." All instructors at the school "are graduates of the School's Human Rights Instructor Certification Course and are prepared to discuss human rights issues as well as integrate human rights and related training" into other courses. "The School," the paper notes, strengthens "the nascent democracies in Latin America" through the "Democratic Sustainment Course which is designed to introduce and teach theory and practice of military and civilian leadership in a constitutional nation-state," and through a course on "Peace Operations" which is designed "to train the student in emerging U.S. doctrine for peace operations strategies."[22]

This rhetoric sounds compelling until we remember that these courses didn't exist for nearly fifty years of the school's life. It is also sobering that El Salvador and Honduras in the 1980s are cited as key SOA success stories, not failures. Colombia is another success story, as is Bolivia, a nation whose dictator, SOA graduate Hugo Banzer, launched a hemisphere-wide effort to persecute progressive church workers.[23]

SOA supporters added courses on human rights and democracy in a cosmetic way in the 1990s and then in public debates claimed these courses reflected the historical mission of the school. The absurdity of doing so is laid bare in the "Historical Edition" of *ADELANTE,* the official publication of the school's commandant.

The final edition of *ADELANTE* included a reprint of a front-page article in a Panamanian newspaper dated April 6, 1961. The headline reads, "U.S. To Set Up Guerrilla War School In C.Z. [Canal Zone]." The opening paragraphs of the article read:

Washington, April 5 (AP) — The U.S. Army said today a special guerrilla and anti-guerrilla warfare school will be set up in the Panama Canal Zone this summer to instruct military personnel of Latin American nations which ask for such training. The school will be established at Ft. Gulick, near a long established jungle warfare training center operated by the army at Ft. Sherman. None of the training at the new school will be designed specifically for any one country, the Army said. There was no immediate explanation of this. It may have been directed at quieting any belief that anti-Castro refugees from Cuba would receive U.S. training at the jungle-guerrilla warfare schools. Classes at the new school will involve guerrilla and anti-guerrilla warfare, intelligence and counter-intelligence, psychological operations, civil affairs and related fields.[24]

Apart from the mention of "civil affairs," the absence of language concerning freedom, democracy, or human rights is rather striking in contrast to images flowing from SOA/WHISC supporters. It is clear that the SOA, in its various forms and manifestations, is primarily a counterinsurgency school.

An additional problem with human rights and democracy propaganda is that the overwhelming majority of courses at the SOA/WHISC remain focused on combat and counterinsurgency. When I debated Colonel Weidner at the Cleveland City Club in March 2000, he mentioned the SOA's "Train the Trainer Course" and stated that "our human rights training is more extensive than at any other DOD [Department of Defense] school." I responded that "only five of the 33 courses offered narrowly relate to human rights. The SOA," I said, "made much of a new course called Human Rights Train the Trainer. The problem was that nobody took the course in 1997 or 1998, and there were no students registered in 1999." Colonel Weidner responded with silence.

A comparative review of SOA and WHISC curricula reveals much the same thing with few substantive changes and a good deal of repackaging. Of thirty-seven courses listed for WHISC, twenty-nine have either exactly the same or only a slight variation in title; four WHISC courses with new titles correspond to courses taught at the SOA; of six courses taught previously at the SOA but not listed as part of the WHISC curriculum, four have clear correlations to courses currently offered at WHISC; and, only four of the courses offered are new to WHISC.[25]

More troubling is the fact that the repackaging is intentionally deceptive. For example, as part of a previous "reform" the Psychological Operations course was eliminated. At the same time, a "new" course, "Information Operations," was added. It is virtually identical to the "eliminated" course.[26] A similar tactic was at play when the Psychological Operations curriculum was positioned within a course with the friendly sounding name "Peace Operations." Former SOA instructor Joseph Blair reviewed SOA Watch's comparative assessment of SOA/WHISC curricula and said: "Your understanding of what they did is accurate. I don't see a dime's worth of difference in the two sets of courses. They are the same — identical."[27] Surprisingly, Colonel Glen Weidner, commandant of the SOA leading up to the name change, agrees. When Father Roy Bourgeois asked him why we can't get a course catalog, even after classes at the WHISC have started, he said: "It's not ready yet. Just use the old one. They're basically the same."[28]

Joseph Blair described other similarities between the "old" SOA and the "new" WHISC in an article in the *Columbus Ledger-Inquirer:*

> [T]he congressional legislation that renamed SOA the Western Hemisphere Institute for Security Cooperation can be likened to Exxon's renaming their Valdez oil tanker in order to appease outraged environmentalists. Unfortunately, although spilled oil can be cleaned up, the suffering and memories of blood, murder, rape, torture and oppression at the hands of Latin American SOA-trained armies will endure forever.... An objective review of WHISC clearly shows that not only has nothing of note changed for the good but in fact, the new SOA will have substantially fewer restrictions and control over how it operates in the future. In the critical area of instructional curriculum, WHISC will not only teach everything SOA did, it now has the added open-ended mission to teach "any other matter the Secretary of Defense determines appropriate."
> ... [T]he new WHISC legislation allows DOD to pay its fixed cost from "any funds" from fiscal year appropriations.... Like SOA, WHISC continues to be commanded by a U.S. Army colonel and the commanding general of Fort Benning and has essentially the same powerless advisory Board of Visitors that has a limited scope of responsibilities.... The new WHISC legislation has many glaring omissions. It fails to establish criteria for selecting foreign students. It places no policy limits on how the institute will be used to promote sales of U.S. military equipment. It has no limits on teaching state-of-the-art military technology. It places no limits on WHISC to man mobile training teams that would operate in Latin America...WHISC, like the SOA, will continue to function as a quasi-Latin American army school since the majority of its ranking faculty members will be foreign officers, and its deputy

director or commandant is also a non-U.S. officer. . . . The current U.S. military initiative in South America, Plan Colombia, that will spend $1.6 billion to fight the losing drug war in Colombia appears now to be logically linked to WHISC's mission and future. Weekly reports of kidnapping, brutal murders, and paramilitary oppression in Colombian villages routinely describe apparent army uniformed personnel involvement. Where is the evidence in Colombia of SOA's past successes from its human rights, democracy, and military professionalization training?[29]

The "new school" language and human rights and democracy rhetoric used by SOA/WHISC defenders is primarily for propaganda purposes. What may be less clear, however, is how such rhetoric fits in with what Colonel Haro Ayerve calls unipolarity. Unipolarity, globalization, and foreign policy implemented through the Golden Straitjacket, the WTO, NAFTA, the IMF, and the Electronic Herd profoundly shape the present mission of SOA/WHISC. The school's strategic role in the context of the economic fist is that it helps Latin American soldiers discern when repressive force is necessary and when it is counterproductive.

The economic fist presents U.S. foreign policy with numerous needs and options. Human rights are important in some situations and expendable in others. One must always explore the possibility that the economic fist can achieve desired outcomes as efficiently and more subtly than dictators or death squads. Rhetoric and courses on human rights and democracy, therefore, respond to critics and to new opportunities posed by the reality of power projection through economic leverage made possible in the context of globalization.

The SOA/WHISC mission today includes helping Latin American officers and soldiers discern when objectives can be achieved through economic rather than military means. Globalization increases instruments of power associated with the economic fist and although the economic fist isn't carried out directly by the school's graduates, studying the possibilities and limitations of new methods of power projection is now featured in the SOA/WHISC curriculum.

The propaganda and discernment roles assigned to the SOA/ WHISC and its graduates, like corporate-led globalization itself, lead back inevitably to the school's repressive role. "Without America on duty," Friedman writes, "there will be no America on line." "The globalization era may well turn out to be the great age of civil wars, not interstate wars."[30] As policies central to the economic fist, the Golden Straitjacket and the Electronic Herd cause enormous human suffering, and as people grow frustrated, organize or rebel, policy makers will be ready. They will revert back to repressive options most common in Stages 1 and 2 of U.S. foreign policy. Without fundamental changes in U.S. foreign policy and in the corporate-dominated global system, we will be headed for many "new El Salvadors" and "new Colombias." The SOA/WHISC and its graduates have more than fifty years experience in repressive tactics. This is why the Pentagon and the White House fight so hard to keep the school open. And this is why closing the SOA/WHISC remains a vital issue and why it must be part of broad-based movements to change U.S. foreign policy, and to promote democratic alternatives to corporate-dominated globalization.

There is another important reason to close the SOA/WHISC. By keeping the SOA open through the ruse of a name change, Pentagon leaders and other SOA supporters in the White House and Congress send a calculated anti-human rights message. This message is clearer and more powerful than any human rights rhetoric or democratic enhancement course offered at the SOA/WHISC: what has changed is geopolitics, not the school; the any means philosophy and strategy at the heart of U.S. foreign policy remains in place.

There has been no repentance, no confession, no expression of remorse, no acknowledgement of wrongdoing, no restitution, and no accountability. What we have is the same school with a different name serving similar interests in a changed geopolitical context. This amounts to a green light for repressive tactics whenever they are deemed necessary. It sends a signal to U.S. leaders, military trainers, and CIA officials that crimes committed in the conduct of foreign policy are acceptable and that the perpetrators of such crimes are immune from prosecution. It signals to Latin American soldiers that repressive tactics, in opposition to what official rhetoric might say, are considered valuable, vital, and necessary so long as they promote U.S. interests. They can be used and will be covered up. This message, the school, and the foreign policy behind it do not bode well for the people of the United States or Latin America.

More than any other nation in the world the United States needs a Truth Commission and much of the world knows it. "In a move that reflected a growing frustration with America's attitude toward international organizations and treaties," Barbara Crossette wrote in the *New York Times,* "the United States was voted off the United Nations Human Rights Commission today for the first time since the panel's founding under American leadership in 1947."[31] This action does not signify a major human rights victory because, as Joanna Weschler of Human Rights Watch noted, the commission members today constitute "a rogues' gallery of human rights abusers."[32] One might say that the U.S. ouster reduces the rogue gallery by one. The exclusion of the United States was clearly warranted, as the U.S. branch of Amnesty International noted on its fortieth anniversary. "We have no prominent leaders in government sounding the clarion call for human rights," executive director William Schulz lamented. "It is no wonder that the U.S. was ousted from the United Nations Human Rights Commission," he said. "That defeat was precipitated by waning U.S. influence and double standards practiced by various administrations and Congresses."[33]

Schulz cited the U.S. failure to ratify a convention to ban anti-personnel land mines and opposition to the establishment of an international court of justice as examples of failed policies. The reasons for U.S. refusal to support an international court of justice should be clear to anyone reading this book with an open mind. There must be numerous fingerprints and tracks left behind when CIA and SOA training manuals advocated torture and terror, SOA graduates targeted progressive religious for a wave of repression, Secretary of State Kissinger arranged a hit on a Chilean general guilty of upholding democracy, the U.S. supported dictatorships and bloody coups throughout Latin America, trained foreign intelligence agents and SOA graduates at the center of numerous disappearances and massacres, and others left behind in the context of other strategic and tactical excesses at the heart of the "any means necessary foreign policy," both past and present.

A *Minneapolis Star Tribune* article, "Nations await U.S. signal on permanent war crimes tribunal: Exemptions for Americans sought," makes it clear that the lack of U.S. support for the tribunal is rooted in Pentagon anxieties. The article notes that "objections from the Pentagon have forced the [Clinton] administration to demand a guarantee that no U.S. officer or civilian official on duty abroad will fall under its jurisdiction, even if the United States is not a party to the court."[34]

Closing the SOA/WHISC is essential to any measure of decency. It is, however, an important but limited goal. The SOA movement is part of a broader movement to change U.S. foreign policy and to revitalize democracy by infusing the political system and electoral politics with the vitality, wisdom, and direction flowing from a politics of nonviolent protest. People's experiences conflict sharply with the new mythology that corporate-led globalization is good for everyone and driven by a "benign hegemon." There is good news to report. Worldwide, people are taking to the streets because the constricted democracy promoted within the corporate-led global system undermines authentic democracy, confines electoral politics in the prison of moneyed interests, and limits economic and political decision-making to the privileged few. Across the nation and the world we are organizing alternatives and demanding change.

Many thousands of us are committed to closing the SOA/WHISC and to changing the foreign policy behind it. We are also working for deeper debt relief, reform or elimination of the IMF-World Bank structural adjustment programs, greater justice and equality in the domestic and global economy, reduced military spending, elimination of the WTO (perhaps the most undemocratic of all global institutions), and environmental justice. We are, in my view, in the early stages of a new movement that has the potential to respond creatively and effectively to the world's most pressing problems (all aggravated by corporate-led globalization). Our important agendas involve both protest of present injustices and the construction of alternative institutions rooted in visions of justice and environmentally sustainable societies.

Changes, including closing the SOA/WHISC, will not come without struggle. We need to continue to educate ourselves and others, do effective grassroots organizing close to home with our congressional delegations, build alliances, participate in national and international protests at the SOA/WHISC's home at Fort Benning, Georgia, and support efforts to challenge and overturn the abusive undemocratic power of free trade agreements and institutions like the IMF and WTO.

In May 2001 a U.S. judge sentenced twenty-six protesters on trial for actions of faithful obedience at the SOA/WHISC. Most will spend six months in prison. A country that protects and rewards its potential war criminals complicit in the many crimes recounted in this book and that gives significant jail time to people protesting these injustices, people who serve the poor and work compassionately in their communities, is in serious trouble. There is much work to do. Those of us who do it for reasons of faith can find inspiration in the words spoken to the judge by SOA/WHISC prisoner of conscience Gwen Hennessey. Gwen and her eighty-eight-year-old sister Dorothy (both Franciscan Sisters) were given six month prison sentences:

> We are here on earth to build the Reign of God — to live out our humanity, to share the God-giftedness within, in solidarity with all of our brothers and sisters and all of creation. We are here to live out our humanity in love, compassion and all of our God-giftedness. Today we cry out for those who are not afforded their dignity. Today we cry out for the voiceless. Today we cry out against unjust structures that teach oppression, rape, and murder, those structures held up by US tax funds, our corporate sin! In the name of God we must shut down the SOA and all that it stands for. We are here on earth to build the Reign of God.

Organizational Descriptions

Free Trade Area of the Americas (FTAA): At the first Summit of the Americas meeting (December 1994), 34 heads of state in attendance from nations in the Western Hemisphere pledged to create a Free Trade Area of the Americas by 2005. The FTAA would eliminate investment and trade barriers on nearly all goods and services traded by member countries. The Summit of the Americas meeting in Canada, April 2001, pushed the FTAA agenda forward and was the subject of mass protests.

International Monetary Fund (IMF): The IMF was created in 1944 in anticipation of currency problems that nations would face following World War II. The IMF was designed to help countries manage balance of payments difficulties. Voting power in the IMF is determined by the size of a nation's contribution to the fund. In the 1980s the IMF worked in conjunction with powerful private banks and governments to manage the third world debt crisis. It imposed conditions (structural adjustment programs) on third world country governments seeking loans from international or private sources. It continues this role today, saying that the conditions it advocates are necessary for third world governments to improve economic management.

North American Free Trade Agreement (NAFTA): NAFTA is a trade agreement between the United States, Mexico, and Canada designed to increase trade and investment. NAFTA seeks a radical reduction in tariff and non-tariff barriers and establishes comprehensive rules governing trade and investment. Labor and environmental concerns are relegated to side agreements with few enforcement provisions. NAFTA took effect on January 1, 1994. The Zapatista movement in Chiapas launched military operations on this day in the southern Mexican state of Chiapas to call attention to NAFTA's disastrous consequences for indigenous people and Mexican agriculture. The FTAA would extend many of the provisions of NAFTA throughout the Western Hemisphere.

World Bank (WB): The WB was created in 1944 along with the IMF. World Bank loans were aimed at helping third world countries develop the infrastructure base deemed necessary for development. Historically, WB projects have reflected biases of Western economic elites that big projects equal or lead to development measured by increases in trade and economic activity rather than in improvements in social well-being. Throughout the 1980s and 1990s, the WB imposed structural adjustment programs on poor countries nearly identical to those of the IMF. Today, there are some cracks in the IMF-WB alliance. The WB is supposed to address issues of poverty. Some voices in the WB are now concerned that social breakdown in many poor countries, perhaps aggravated by IMF conditionality, may make these countries ungovernable.

World Trade Organization (WTO): The WTO emerged out of a series of international negotiations to set rules concerning international trade and investment under the auspices of the General Agreement on Tariffs and Trade (GATT). The WTO was created in 1994. It oversees and has a dispute-settlement mechanism to enforce rules of international trade. The WTO is perhaps the most powerful, antidemocratic organization in the world, with dispute panels made up of unelected, non-accountable trade attorneys. It is located in Geneva, Switzerland. Massive protests at the WTO meeting in Seattle (November–December 1999) brought workers, farmers, students and environmentalists together in an impressive demonstration of people's democracy. The *Oxford Analytica Daily Brief* on World Trade dated December 4, 2000, notes that the "failed Ministerial Conference in Seattle inflicted lasting damage on the WTO. Much of the past year," the report says, "has been characterized by a process of sustained convalescence, which will continue in 2001" (p. 7).

Notes

Introduction

1. I address the U.S.-sponsored war against liberation theology in a novel recently published by EPICA. See Jack Nelson-Pallmeyer, *Harvest of Cain* (Washington, D.C.: EPICA, 2001; www.epica.org).

2. By military-industrial-congressional complex I mean to implicate all the powerful groups that promote and benefit from excessive military spending. These groups include the military branches themselves, the businesses that produce the weapons, and congressional representatives whose districts or states are the sites of unneeded bases or defense industry plants.

3. David C. Korten, *When Corporations Rule the World* (West Hartford, CT: Kumarian Press, and San Francisco: Berret-Koehler Publishers, 1995), pp. 261–62.

4. "Discoveries Underline Need for Truth Commission," *National Catholic Reporter*, May 19, 2000.

5. *ADELANTE: U.S. Army School of the Americas 1946–2000*, pp. 2–3. *ADELANTE* is the official publication of the SOA's commandant.

Chapter 1: Official History and the People's Stories

1. *The CIA's Nicaraguan Manual: Psychological Operations in Guerrilla Warfare* (New York: Random House, 1985), p. 33.

2. "Affidavit of Edgar Chamorro," Case Concerning Military and Paramilitary Activities in and against Nicaragua (Nicaragua *v.* United States of America), International Court of Justice, September 5, 1985, p. 21.

3. William I. Robinson and Kent Norsworthy, *David and Goliath:*

The U.S. War against Nicaragua (New York: Monthly Review Press, 1987), pp. 56–57, emphasis added.

4. Quoted in Leslie Cockburn, *Out of Control* (New York: Atlantic Monthly Press, 1987).

5. "Affidavit of Edgar Chamorro," pp. 20–21.

6. See Jack Nelson-Pallmeyer, *War against the Poor: Low-Intensity Conflict and Christian Faith* (Maryknoll, NY: Orbis Books, 1989), chapter 3.

7. Bob Woodward, *Veil: The Secret Wars of the CIA* (New York: Simon & Schuster, 1987), pp. 195 and 173.

8. Robinson and Norsworthy, *David and Goliath*, p. 26.

9. "Affidavit of Edgar Chamorro," p. 17.

10. "Report on the Guatemala Review," Intelligence Oversight Board, June 28, 1996, p. 84.

11. Lisa Haugaard, "Torture 101," *In These Times*, October 14, 1996, p. 14.

12. Dana Priest, "U.S. Instructed Latins on Executions, Torture," the *Washington Post*, September 21, 1966.

13. "Torture 101," *In These Times*, p. 14.

14. Arthur Jones, "Haiti, Salvador Links Viewed," *National Catholic Reporter*, November 19, 1993.

15. "The Ties That Bind: Colombia and Military-Paramilitary Links," February 2000, Vol. 12, No. 1 (B), from the web at www.hrw.org/reports/2000/colombia.

16. Greg Gordon, "D.C. Diary," *Minneapolis Star Tribune*, June 1, 2001.

17. Americas Watch testimony, January 31, 1990, emphasis added.

18. Celerino Castillo III and Dave Harmon, *Powderburns: Cocaine, Contras and the Drug War* (Buffalo, NY: Oakville and London: Mosaic Press, 1994), pp. 151–54.

19. "U.S. Army School of the Americas Frequently Asked Questions," emphasis added.

20. Adam Isacson and Joy Olson, *Just the Facts 1999 Edition: A Civilian's Guide to U.S. Defense and Security Assistance to Latin America and the Caribbean* (Washington, D.C.: Latin America Working Group, 1999), p. ix.

Chapter 2: Guns, Greed, and Globalization: Continuity and Change

1. Quoted in an article by Michael Klare, "Low Intensity Conflict: The War of the 'Haves' against the 'Have-Nots,'" *Christianity and Crisis,* February 1, 1988, p. 12.

2. Quoted in *Sojourners,* February–March 1990, p. 5.

3. See Jack Nelson-Pallmeyer, *War against the Poor: Low-Intensity Conflict and Christian Faith* (Maryknoll, NY: Orbis Books, 1989).

4. As quoted by Klare in *Christianity and Crisis,* February 1, 1988, pp. 12–13.

5. Statistics are from the United Nations Development Program's 1998 *Human Development Report,* as quoted in Chuck Collins et al., *Shifting Fortunes: The Perils of the Growing American Wealth Gap* (Boston: United For a Fair Economy, 1999), p. 18.

6. Thomas L. Friedman, *The Lexus and the Olive Tree: Understanding Globalization* (New York: Farrar, Straus and Giroux, 1999), pp. 12 and 197.

7. Ibid., emphasis in original.

8. Ibid., p. 12.

9. Ibid., p. 376.

10. Ibid., p. 373.

11. Quoted in Richard Barnet, *Intervention and Revolution* (New York: World Publishing Company, 1969), pp. 229–30.

12. Adam Isacson and Joy Olson, *Just the Facts 1999 Edition: A Civilian's Guide to U.S. Defense and Security Assistance to Latin America and the Caribbean,* Washington, D.C., a project of the Latin American Working Group in cooperation with the Center for International Policy, 1999, p. iv, emphasis (bold face) in original.

Chapter 3: Focus on the SOA

1. Howard Zinn, *A People's History of the United States* (New York: HarperCollins, 1999), p. 408.

2. "A Half Century of Professionalism: The U.S. Army School of the Americas" by Joseph C. Leuer, in the "Historical Edition" of

ADELANTE: U.S. Army School of the Americas 1946–2000, p. 6. The magazine says, "Questions concerning material should be addressed to the U.S. Army School of the Americas, Public Affairs Office, Fort Benning, GA 31905."

3. Under the terms of the 1977 Canal Treaty the SOA had to leave Panama by the end of 1985.

4. Vincent Harding, "We Must Keep Going: Martin Luther King Jr. and the Future of America," in Walter Wink, ed., *Peace Is the Way: Writings on Nonviolence from the Fellowship of Reconciliation* (Maryknoll, NY: Orbis Books, 2000), pp. 194–95.

5. Ibid., p. 198.

6. Ibid.

7. Ibid., p. 199.

8. Ibid., p. 198.

9. Ibid., p. 201.

10. A letter sent to "His Excellency, The President of the United States, Mr. Jimmy Carter" on February 17, 1980. In Archbishop Oscar Romero, *Voice of the Voiceless* (Maryknoll, NY: Orbis Books, 1985), pp. 188–90.

11. Quoted in an article by Michael Klare, "Low Intensity Conflict: The War of the 'Haves' against the 'Have-Nots,'" *Christianity and Crisis,* February 1, 1988, pp. 12–13.

12. Martin Lange and Reinhold Iblacker, eds., *Witnesses of Hope: The Persecution of Christians in Latin America* (Maryknoll, NY: Orbis Books, 1981), pp. 79–80.

13. Michael K. Frisby, "U.S. Aid: Rebels Gained in Priests Killing," *Boston Globe,* December 20, 1989, p. 17.

14. *United Nations Truth Commission Report,* March 15, 1993.

15. For more information on SOA Watch and the movement to close the SOA/WHISC contact SOA Watch, PO Box 4566, Washington, DC 20017; 202-234-3440, or www.soaw.org.

16. Col. Glenn R. Weidner, "A Word from the Commandant," in the "Historical Edition" of *ADELANTE: U.S. Army School of the Americas 1946–2000,* p. 2.

17. For clarity I refer to the School in its various forms from 1946 to the time of the most recent name change as the School of the Americas,

or SOA. I refer to the School as SOA/WHISC to indicate the School's status following the SOA's "closure" in December 2000 and reopening as the Western Hemisphere Institute for Security Cooperation (WHISC) in January 2001.

18. Weidner, in the "Historical Edition" of *ADELANTE*, p. 6.

19. Ibid., emphasis added.

Chapter 4: Evidence and Tactics

1. *United Nations Truth Commission Report*, March 15, 1993.

2. "U.S., Latin America Sign Secret Defense Plan," *National Catholic Reporter*, December 16, 1988.

3. Quoted in *Total War against the Poor* (New York: Circus Publications, 1990), p. 133.

4. As quoted in Jon Sobrino, Ignacio Ellacuría and Others, *Companions of Jesus: The Jesuit Martyrs of El Salvador* (Maryknoll, NY: Orbis Books, 1990), p. xviii.

5. The Committee of Santa Fe, "A New Inter-American Policy for the Eighties" (Washington, D.C.: Council for Inter-American Security, 1980).

6. Quoted in *Guatemala: Never Again! Recovery of Historical Memory Project, The Official Report of the Human Rights Office, Archdiocese of Guatemala* (Maryknoll, NY: Orbis Books, 1999), pp. xxviii–xxix.

7. Ibid., pp. 152 and xvi.

8. I have written in detail elsewhere about management of terror as part of U.S. psychological warfare tactics within the framework of low-intensity-conflict strategy. See Jack Nelson-Pallmeyer, *War against the Poor: Low-Intensity Conflict and Christian Faith* (Maryknoll, NY: Orbis Books, 1989). I have also written about the crucifixion of Jesus in the context of Roman psychological terror in first century Palestine. See Jack Nelson-Pallmeyer, *Jesus against Christianity: Reclaiming the Missing Jesus* (Harrisburg, PA: Trinity Press International, 2001).

9. "Lessons in Terror," *Boston Globe*, October 1, 1996.

10. *Guatemala: Never Again!*, 118–19.

11. Ibid., p. 120.

12. Ibid., pp. 123–24.

13. Ibid., p. 36, emphasis in original.

14. Ibid., p. 105.

15. Ibid., p. 155.

16. Ibid.

17. Ibid.

18. Ibid.

19. "Our Man in Guatemala," *Washington Post*, March 26, 1995.

20. Ibid.

21. Tim Weiner, "A Guatemalan Officer and the CIA," *New York Times*, March 26, 1995.

22. "The Vigil Begins," excerpted from *Sojourners*, July–August, 1996, p. 18.

23. Ibid., pp. 18–19.

24. General Gramajo is featured centrally within the *Guatemala: Never Again!* report as a representative of a military faction attempting to transition the Guatemalan armed forces into a postwar role. Under U.S. tutelage, his challenge in the late 1980s was to move the Guatemalan military away from its historic role as an instrument of indiscriminate terror to one of calculated violence within the framework of U.S. low-intensity-conflict strategy (see p. 252). His favorable treatment by U.S. leaders in the context of his role as head of the armed forces at the time of the rape and torture of Sister Diana Ortiz is another example of U.S. complicity in religious persecution.

25. Daniel Maloney, "SOA Recognizes 1991 Staff College Graduates," *The Bayonet*, January 3, 1992.

26. *Guatemala: Never Again!*, pp. xxiii–xxiv.

27. Ibid., pp. xxiv–xxv, emphasis in original.

Chapter 5: More Evidence and Key Questions

1. Gary Cohn and Ginger Thompson, "Unearthed: Fatal Secrets," *Baltimore Sun*, reprint of a series that appeared June 11–18, 1995.

2. *Inside the School of Assassins*, a documentary film produced by Robert Richter, Richter Productions. This hour-long documentary can be ordered from SOA Watch.

3. Ibid.

4. *Baltimore Sun,* June 11–18, 1995.

5. Ibid.

6. See a commentary by Joseph E. Mulligan, "What Did Negroponte Hide and When Did He Hide It," *Los Angeles Times,* April 19, 2001.

7. Ibid.

8. Christopher Hitchens, "The Case against Henry Kissinger, Part One: The Making of a War Criminal," *Harper's Magazine,* February 2001, p. 53

9. Ibid., pp. 53–54.

10. Ibid., p. 55.

11. Christopher Hitchens, "The Case against Henry Kissinger, Part Two: Crimes against Humanity," *Harper's Magazine,* March 2001, p. 50, emphasis added.

12. Ibid., p. 52.

13. "The Case against Henry Kissinger," Part One, p. 36.

14. Javier Giraldo, S J , *Colombia: The Genocidal Democracy* (Monroe, ME: Common Courage Press, 1996), pp. 19–20.

15. Ibid., pp. 7–8.

16. Ibid., p. 46.

17. Ibid., p. 60.

18. Quoted in "Plan Colombia: Wrong Issue, Wrong Enemy, Wrong Country," by Marc Cooper, *The Nation,* March 19, 2001, p. 16.

19. "The Ties That Bind: Colombia and Military-Paramilitary Links," Human Rights Watch, February 2000, Vol. 12, No. 1 (B), at www.hrw.org/reports/2000/colombia.

20. The following information is provided by SOA Watch and is based on a variety of human rights reports.

21. As quoted in information from SOA Watch.

22. Witness for Peace Delegation to Colombia, January 5–17, 2001, as reported via email by Gail Phares.

23. www.ciponline.org

24. From the rough transcript of an interview conducted by Rose Berger in Bogotá, Colombia, as part of a Witness for Peace delegation, January 2001.

25. Winifred Tate, "Repeating Past Mistakes: Aiding Counter-insurgency in Colombia," *NACLA Report on the Americas,* September–October, 2000, p. 17.

26. "Adjusting Drug Policy," *New York Times,* February 27, 2001.

27. "Repeating Past Mistakes," p. 18.

28. "Plan Colombia: Wrong Issue, Wrong Enemy, Wrong Country," *The Nation,* March 19, 2001, p. 11.

29. Ibid., p. 17.

30. *Congressional Record,* May 20, 1994, p. H3771.

31. Ibid.

Chapter 6: Geopolitics and the SOA/WHISC: Foreign Policy Stage 1

1. Speech before the National Foreign Trade Convention, November 12, 1946.

2. Quoted in Michael T. Klare and Peter Kornbluth, eds., *Low Intensity Warfare: Counterinsurgency, Proinsurgency, and Antiterrorism in the Eighties* (New York: Pantheon Books, 1988), p. 48.

3. Hubert Humphrey, 84th Congress, First Session, Senate Committee on Agriculture and Forestry, *Hearings: Policies and Operations of Public Law 480,* 1957, p. 129.

4. Directorate of Intelligence Office of Political Research, *Potential Implications of Trends in World Population, Food Production, and Climate* (Washington, D.C.: Library of Congress, 1974), pp. 15 and 39.

5. *Agribusiness Manual* (New York: The Interfaith Center on Corporate Responsibility, 1978), Section II, p. 13.

6. From the rough transcript of an interview conducted by Rose Berger in Bogotá, Colombia, as part of a Witness for Peace delegation, January 2001.

7. "School of the Americas and U.S. Foreign Policy Attainment in Latin America," an "information paper" by Joseph C. Leuer, January 1996, p. 7, emphasis added.

8. Ibid., p. 1.

9. Ibid.

10. Ibid., p. 2, emphasis added.

11. Ibid., pp. 2–3.

12. *Atlanta Constitution,* June 3, 1995.

13. This quotation is taken from the written transcript of a Public Affairs Television special with Bill Moyers, entitled *The Secret Government: The Constitution in Crisis.* The program was a production of Alvin H. Perlmutter, Inc., and Public Affairs Television, Inc., in association with WNET and WETA. Copyright 1987 by Alvin H. Perlmutter, Inc., Public Affairs Television, Inc. The written transcript was produced by Journal Graphics, Inc., New York.

14. Thomas McCann, *An American Company: The Tragedy of United Fruit* (New York: Crown Publishers, 1976), pp. 39–40.

15. Peter Dale Scott, Jonathan Marshal, and Jane Hunter, *The Iran Contra Connection: Secret Teams and Covert Operations in the Reagan Era* (Boston: South End Press, 1987), p. 31, emphasis in original.

16. José Comblin, *The Church and the National Security State* (Maryknoll, NY: Orbis Books, 1979), p. 65.

17. Penny Lernoux, *Cry of the People* (New York: Penguin Books, 1980), pp. 142–43.

18. Quoted in *Sojourners Magazine,* February–March, 1990.

19. Quoted in Suzanne Gowan et al., *Moving toward a New Society* (Philadelphia: New Society Press, 1976), pp. 86–87.

Chapter 7: Geopolitics and the SOA/WHISC: Foreign Policy Stages 2–4

1. See Jack Nelson-Pallmeyer, *War against the Poor.*

2. *San Antonio Express-News,* April 14, 1995.

3. *Des Moines Register,* May 16, 1995.

4. *Atlanta Constitution,* June 3, 1995.

5. *Cleveland Plain Dealer,* July 20, 1995.

6. *New York Times,* March 24, 1995.

7. Ecumenical Coalition for Economic Justice, *Recolonization or Liberation: The Bonds of Structural Adjustment and Struggles for Emancipation* (Toronto: Ecumenical Coalition for Economic Justice, 1990), p. 6.

8. Ibid., pp. 7, 8 and 24.

9. Michael Harrington, *Socialism Past and Future* (New York: Arcade Publishing, 1989), p. 165.

10. See UNICEF, *The State of the World's Children 1989* and *The State of the World's Children 1990* (New York: Oxford University Press, 1989 and 1990).

11. *Recolonization or Liberation*, p. 30.

12. Quoted in Walden Bellow, *Brave New Third World? Strategies for Survival in the Global Economy* (San Francisco: The Institute for Food and Development Policy, 1989), pp. 60–61.

13. Thomas L. Friedman, *The Lexus and the Olive Tree: Understanding Globalization* (New York: Farrar, Straus and Giroux, 1999), p. 8.

14. Ibid., p. 11.

15. Ibid., p. 87.

16. Ibid., p. 88.

17. John B. Cobb, Jr., "The Theological Stake in Globalization," a talk given in Northfield, Minnesota, Summer 2000.

18. *The Lexus and the Olive Tree*, p. 116.

19. Ibid., p. 142.

20. Ibid., p. 86.

21. Ibid., p. 350.

22. Ibid., p. 374.

23. Ibid., p. 17, emphasis added.

24. Ibid., pp. 212–13.

25. Col. Patricio Haro Ayerve, "Greetings from the Subcommandant," quoted in *ADELANTE: U.S. Army School of the Americas 1946–2000*, p. 4.

26. Ibid., p. 204.

27. Larry Rohter, "Latin America's Armies Are Down but Not Out," *New York Times*, June 20, 1999.

28. *The Lexus and the Olive Tree*, p. 29.

29. Ibid., p. 373.

30. *Marine Corps Gazette*, May 1990, p. 16.

31. Quoted in Michael Klare, "Facing South: The Pentagon and the Third World in the 1990s," a talk given at the University of Minnesota, October 5, 1990.

32. James Petras, "The Meaning of the New World Order: A Critique," *America*, May 11, 1991, p. 512.

33. Andrew and Leslie Cockburn, *Dangerous Liaison: The Inside Story of the U.S.-Israeli Covert Relationship* (New York: HarperCollins, 1991), p. 354–55.

34. Ibid., p. 355.

35. Coletta Youngers, "Cocaine Madness: Counternarcotics and Militarization in the Andes," *NACLA Report on the Americas*, November–December 2000, p. 18.

36. *New York Times*, June 20, 1999.

37. Cedric Muhammad, "A Deeper Look: Concerned about Cheney," http://blackelectorate.com/archives/072500.asp.

38. See Bill Hartung, "Rumsfeld Reconsidered: An Ideologue in Moderate's Clothing," World Policy Institute, on the website hartung@newschool.edu.

39. William D. Hartung, "Winning One for the Gipper: Donald Rumsfeld and the Return to the Star Wars Lobby," www.fpif.org/commentary/0101starwars_body.html.

40. William D. Hartung, "Bush's Nuclear Revival," *The Nation*, March 12, 2001, pp. 4–5.

Chapter 8: Globalization and Greed

1. Eric Black, *Rethinking the Cold War* (Minneapolis: Paradigm Press, 1988), pp. 9–11.

2. Thomas L. Friedman, *The Lexus and the Olive Tree: Understanding Globalization* (New York: Farrar, Straus and Giroux, 1999), pp. 6, 8 and 93.

3. Ibid., p. 58.

4. Ibid., p. 50.

5. John B. Cobb, Jr., "The Theological Stake in Globalization," a talk given in Northfield, Minnesota, Summer 2000.

6. *Rethinking the Cold War*, p. 9.

7. *The Lexus and the Olive Tree*, p. 374.

8. Ibid., p. 375.

9. Ibid., p. 304.

10. Ibid., pp. 87 and 88.

11. Ibid., p. 163.

12. Ibid., p. 161.

13. Xabier Gorostiaga, "World Has Become a 'Champagne Glass,'" *National Catholic Reporter,* January 27, 1995.

14. Chuck Collins and Felice Yeskel, *Economic Apartheid in America* (New York: New Press, 2000), p. 61.

15. Quoted by Paul Street, "Free to Be Poor," *Z Magazine,* June 2001, p. 25.

16. Chuck Collins et al., *Shifting Fortunes: The Perils of the Growing American Wealth Gap* (Boston: United for a Fair Economy, 1999), p. 18.

17. *Shifting Fortunes: The Perils of the Growing American Wealth Gap,* pp. 5, 16 and 18.

18. *Economic Apartheid in America,* p. 39, emphasis added.

19. Ibid., p. 58.

20. Ibid., p. 24.

21. Ibid., p. 46.

22. Ibid.

23. Ibid., p. 6.

24. *The Lexus and the Olive Tree,* p. 248.

25. Ibid., p. 249.

26. Ibid., p. 250.

27. Ibid.

28. Ibid., p. 214.

29. *New York Times,* April 24, 2001.

30. "FTAA for Beginners," United for a Fair Economy, January 2001.

31. Witness for Peace, "A Hemisphere for Sale: The Epidemic of Unfair Trade in the Americas," 2001, p. 1, www.witnessforpeace.org. The quoted paragraph contains statistics from Fundo de Apoyo Mutuo, Mexico, 2000 and Grupo Parlamentario, PRD, 2000.

32. *The Lexus and the Olive Tree,* pp. 72 and 74.

33. "A Hemisphere for Sale," p. 1.

34. *Z Magazine,* June 2001, pp. 25–26.

35. *Economic Apartheid in America,* p. 25.

36. Paul Hawken, *The Ecology of Commerce* (New York: Harper Business, 1993), p. 6.

37. David C. Korten, *When Corporations Rule the World* (West Hartford: Kumarian Press and San Francisco: Berret-Koehler Publishers, 1995), p. 261.

38. The study is available at http://www.ips-dc.org/top200.htm

39. *When Corporations Rule the World*, p. 262.

40. Sandra Postel and Christopher Flavin, *State of the World 1991* (New York: W. W. Norton, 1991), p. 174.

41. Ed Ayres, editor, quoted in a "Note from a Worldwatcher," *World Watch*, March–April, 2001, p. 3.

42. Alan Durning, *State of the World 1990* (New York: W. W. Norton, 1990), pp. 135–36.

43. In Lester Brown et al., *State of the World 1994*, p. 19.

44. Rapid population growth is fueled by three factors: (1) many women are illiterate, impoverished, and have almost no access to effective methods of family planning; (2) cultural factors associate large numbers of children with status or divine blessing; and, (3) the world's population is young, which means that there are large numbers of women of child bearing age. This means that the global population would grow significantly over the short and medium term even if all women had only two children. On his first day in office, President George W. Bush cut U.S. support for international family planning.

45. *State of the World 1991*, p. 177.

46. "Intergovernmental Panel on Climate Change: Working Group II, Climate Change Impacts, Adaptation and Vulnerability," Shanghai Draft, January 21, 2001, available on line at www.usgcrp.gov.ipcc.

47. Alan Durning, *How Much Is Enough* (New York: W. W. Norton, 1992), p. 22.

48. *The Lexus and the Olive Tree*, p. 235.

49. Ibid., p. 8.

50. Ibid., p. 221.

51. *How Much Is Enough*, pp. 43 and 46.

52. *When Corporations Rule the World*, pp. 18–19.

53. Quoted in *The Lexus and the Olive Tree*, p. 11.

54. *State of the World 1990,* p. 16.

55. *The Lexus and the Olive Tree,* p. 212, emphasis in original.

Chapter 9: A Rose by Any Other Name

1. For information on the movement to close the SOA/WHISC contact SOA Watch, PO Box 4566, Washington, DC 20017, 202-234-3440, www.soaw.org.

2. "A Half Century of Professionalism: The U.S. Army School of the Americas," by Joseph C. Leuer in the "Historical Edition" of *ADELANTE: U.S. Army School of the Americas 1946–2000,* p. 30.

3. "Still a School of Assassins," as quoted in a flyer produced by SOA Watch.

4. Ibid.

5. Ibid.

6. Bill numbers change and so I ask interested readers to contact SOA Watch at the address noted above.

7. "SOA Quick Fix Won't Cleanse U.S. Policy," *National Catholic Reporter,* June 2, 2000.

8. Col. Glenn R. Weidner, "A Word from the Commandant," *ADELANTE,* p. 2, emphasis added.

9. "New Name...New Game," an interview with Col. Richard D. Downie by Dusty Dix, *Columbus Ledger-Enquirer,* April 29, 2001, emphasis added.

10. "School Has Only Changed Names," by Joseph A. Blair, *Columbus Ledger-Enquirer,* January 24, 2001.

11. "Ft. Benning's New 'Potemkin Village' Masks Its Shame," by Ed Kinane, *Post-Standard,* February 7, 2001, emphasis in original.

12. "A Half Century of Professionalism," pp. 25 and 26.

13. Thomas L. Friedman, *The Lexus and the Olive Tree: Understanding Globalization* (New York: Farrar, Straus and Giroux, 1999), p. 35.

14. Ibid., p. 273.

15. Ibid., pp. 375 and 373.

16. Ibid., p. 362.

17. Ibid., p. 130.

18. Ibid., p. 115.

19. Ayerve, "Greetings from the Subcommandant," in *ADELANTE,* p. 4.

20. Ibid.

21. Editorial by Frank del Olmo, *Los Angeles Times,* April 3, 1995, emphasis in original.

22. "School of the Americas and U.S. Foreign Policy Attainment in Latin America," an "information paper" by Joseph C. Leucr, January 1996, pp. 1, 8–10.

23. Ibid., pp. 11–14.

24. The article is from the Panamanian newspaper the *Star and Herald* as quoted in *ADELANTE,* pp. 10–11.

25. "Course Offerings at the Western Hemisphere Institute for Security Cooperation," prepared by SOA Watch.

26. Ibid.

27. Ibid.

28. Ibid.

29. "School Has Only Changed Names," by Joseph A. Blair, *Columbus Ledger-Enquirer,* January 24, 2001.

30. *The Lexus and the Olive Tree,* pp. 376 and 212.

31. Barbara Crossette, "U.S. Voted Off Rights Panel of the UN. for the First Time," from NYTimes.com, May 4, 2001.

32. Ibid.

33. Associated Press, "U.S. No Longer Leader on Human Rights, Says Amnesty International," *Minneapolis Star Tribune,* May 31, 2001.

34. *Minneapolis Star Tribune,* November 24, 2000.

DATE DUE